SpringerBriefs in Mathematics

T0211984

SpringerBriefs in Mathematics showcases expositions in all areas of mathematics and applied mathematics. Manuscripts presenting new results or a single new result in a classical field, new field, or an emerging topic, applications, or bridges between new results and already published works, are encouraged. The series is intended for mathematicians and applied mathematicians.

For further volumes:
http://www.springer.com/series/10030

Yuval Z. Flicker

Drinfeld Moduli Schemes and Automorphic Forms

The Theory of Elliptic Modules with Applications

Springer

Yuval Z. Flicker
Department of Mathematics
The Ohio State University
Columbus, OH, USA

ISSN 2191-8198   ISSN 2191-8201 (electronic)
ISBN 978-1-4614-5887-6   ISBN 978-1-4614-5888-3 (ebook)
DOI 10.1007/978-1-4614-5888-3
Springer New York Heidelberg Dordrecht London

Library of Congress Control Number: 2012952719

Mathematics Subject Classification (2010): 11F70, 22E35, 22E50, 11G09, 11G20, 11G45, 11S37, 14G10, 11F72, 22E55

Printed on acid-free paper

Springer is part of Springer Science+Business Media (www.springer.com)

CONTENTS

1. INTRODUCTION

Let F be a geometric global field of characteristic $p > 0$, $\mathbb{A}$ its ring of adèles, $G = \mathrm{GL}(r)$ and π an irreducible admissible representation of $G(\mathbb{A})$, namely a $G(\mathbb{A})$-module, over $\mathbb{C}$. Then π is the restricted direct product $\otimes_v \pi_v$ over all places v of F of irreducible admissible $G_v = G(F_v)$-modules π_v. For almost all v the component π_v is unramified. In this case there are nonzero complex numbers $z_{1,v}, \ldots, z_{r,v}$, uniquely determined up to order by π_v and called the *Hecke eigenvalues* of π_v, with the following property: π_v is the unique irreducible unramified subquotient $\pi((z_{i,v}))$ of the G_v-module $I(\mathbf{z}_v) = \mathrm{Ind}(\delta^{1/2}\mathbf{z}_v; B_v, G_v)$ which is normalized induced from the unramified character $\mathbf{z}_v : (b_{ij}) \mapsto \prod_i z_{iv}^{\deg_v(b_{ii})}$ of the upper triangular subgroup B_v of G_v.

The first main theme in this work concerns congruence relations (see below).

The second such theme concerns the following *purity theorem* for cuspidal $G(\mathbb{A})$-modules. Let π be a complex cuspidal $G(\mathbb{A})$-module; it is an irreducible admissible representation π of $G(\mathbb{A})$ which occurs as a direct summand in the representation of $G(\mathbb{A})$ by right translation on the space of complex-valued cuspidal functions on $G(F)\backslash G(\mathbb{A})$. *If π has a cuspidal component and a unitary central character, then the absolute value of each Hecke eigenvalue z_{iv} of almost all unramified components π_v of π is equal to one.* This is Theorem 10.8. Its proof uses neither Deligne's conjecture nor the congruence relations. The purity theorem is a representation theoretic analogue of Ramanujan's conjecture concerning the Hecke eigenvalues (or rather Fourier coefficients) of the cusp form $\Delta(z) = e^{2\pi i z}\prod_1^\infty (1 - e^{2\pi i z n})$ on the upper half plane $\mathrm{Im}(z) > 0$ for the group $\mathrm{SL}(2, \mathbb{Z})$.

The third major theme in this work concerns the higher reciprocity law. Let σ be a continuous r-dimensional ℓ-adic representation $\sigma : W(\overline{F}/F) \to \mathrm{GL}(r, \overline{\mathbb{Q}}_\ell)$ of the Weil group of F, which is *constructible*, namely unramified for almost all v. Equivalently σ is a smooth ℓ-adic sheaf on $\mathrm{Spec}\, F$ which extends to a smooth ℓ-adic sheaf on an open subscheme of the smooth projective curve whose function field is F. For such v the restriction σ_v of σ to the decomposition group $W(\overline{F}_v/F_v)$ at v factorizes through $W(\overline{\mathbb{F}}_v/\mathbb{F}_v) \simeq \mathbb{Z}$, where $\mathbb{F}_v$ is the residue field of F_v. The isomorphism class of σ_v is determined by the eigenvalues $\{u_{i,v} = u_i(\sigma_v); 1 \le i \le r\}$ of the (geometric) Frobenius $\sigma_v(\mathrm{Fr}_v)$. Then we say that such σ and the $G(\mathbb{A})$-module $\pi = \otimes \pi_v$ *correspond* if for almost all v the r-tuple $(u_i(\sigma_v))$ is equal, up to order, to the r-tuple $(z_i(\pi_v))$.

The case of $r = 1$ is class field theory: $W(\overline{F}/F)_{\mathrm{ab}} \simeq \mathbb{A}^\times/F^\times$, which in the local case asserts that $W(\overline{F}_v/F_v)_{\mathrm{ab}} \simeq F_v^\times$, normalized by mapping a geometric Frobenius Fr_v to a local uniformizer π_v in $F_v^\times$. Here $\mathrm{Fr}_v \in W(\overline{F}_v/F_v)$ is any element which maps to the inverse φ^{-1} of the "arithmetic" Frobenius substitution $\varphi : x \mapsto x^{q_v}$, which generates $W(\overline{\mathbb{F}}_v/\mathbb{F}_v)$ and is an automorphism of $\overline{\mathbb{F}}_v$ over $\mathbb{F}_v$. Let ∞ be a fixed place of F.

Y.Z. Flicker, *Drinfeld Moduli Schemes and Automorphic Forms: The Theory of Elliptic Modules with Applications*, SpringerBriefs in Mathematics, DOI 10.1007/978-1-4614-5888-3_1, © Yuval Z. Flicker 2013

The special case of the *higher reciprocity law* which is proven in this work asserts that *the correspondence defines a bijection between the sets of* (1) *equivalence classes of cuspidal representations* π *of* $\mathrm{GL}(r, \mathbb{A})$ *whose component* π_∞ *is cuspidal and* (2) *equivalence classes of irreducible r-dimensional continuous ℓ-adic constructible representations* σ *of* $W(\overline{F}/F)$, *or irreducible rank r smooth ℓ-adic sheaves on* $\mathrm{Spec}\, F$ *which extend to smooth sheaves on an open subscheme of the curve underlying F, whose restriction* σ_∞ *to* $W(\overline{F}_\infty/F_\infty)$ *is irreducible.*

This reciprocity law is reduced in Chaps. 11 and 12 to Deligne's conjecture (Theorem 6.8), introduced by Deligne in the 1970s for the purpose of this reduction. The fixed point formula expresses a trace on ℓ-adic cohomology with compact supports and coefficients in a smooth sheaf, in terms of traces on the stalks of the sheaf at fixed points. Deligne's conjecture asserts that the fixed point formula remains valid in the context of the composition of a certain correspondence on a separated scheme of finite type over a finite field and a sufficiently high power, depending on the correspondence, of the Frobenius morphism.

This work existed as an unpublished manuscript since 1983. It is the first openly circulated manuscript where Deligne's conjecture appeared. Its results, first discussed in a seminar with Kazhdan in 1982, were announced in the publication [FK3]. Perhaps this work contributed a little to the interest in Deligne's important conjecture. This was the original purpose of this work, to motivate Deligne's conjecture by means of exhibiting some of its applications. Since then this conjecture was proven unconditionally by Fujiwara [Fu] in 1997 and later by Varshavsky [V] in 2007, after work by Zink [Z], Ed Shpiz [Sh], and Pink [P] in special cases. This completed that part of our work which relied on Deligne's conjecture. Our results concern the full local correspondence (for $\mathrm{GL}(r)$ over a local field of positive characteristic) as well as the global correspondence for cuspidal representations (of $\mathrm{GL}(r, \mathbb{A})$ over a global function field) which have a cuspidal component. They are based on Drinfeld's theory of elliptic modules, also named Drinfeld modules by Deligne, introduced by Drinfeld [D1], [D2] in 1974 and 1977 to prove the reciprocity law when $r = 2$.

Drinfeld later introduced a generalization, which he named Shtukas, to remove the restriction that the global cuspidal representations have a square-integrable component. The work was carried out by Lafforgue [Lf1] and [Lf2] in 1997 and 2002, who in addition to Deligne's (proven) conjecture used the full trace formula of Arthur in the function field case, to obtain the reciprocity law for any global cuspidal representation. His important work is nevertheless technically very challenging, so on the occasion of teaching a course on the topic at OSU in winter 2012, we updated our work to include the references to the proofs of Deligne's conjecture and other works that continued and extended ours. I hope it is still of interest not only as the first work where the local correspondence and a major case of the global correspondence were established in the general rank case but also since our work is considerably simpler than that of Lafforgue, as we use only a simple case of the trace formula, and the relatively elementary theory of elliptic (= Drinfeld) modules.

In particular we were led to develop in Chap. 13 a "simple" converse theorem for $GL(r)$ over a function field, "simple" meaning for cuspidal global representations with cuspidal local components at a fixed finite set of places of the global field F. However, note that the converse theorem is not used in the decomposition of the cohomology (see below), neither in the proof of the existence theorem (for each π there is a σ) nor in the proof of the local correspondence. It is used only to show the surjectivity of the map $\pi \mapsto \sigma$.

Thus, assuming Deligne's conjecture (Theorem 6.8), we show in Chap. 12 that there exists a unique bijection, denoted $\pi_v \mapsto \sigma_v$ or $\sigma_v \mapsto \pi_v$ and called the *local reciprocity correspondence*, between the sets of (1) equivalence classes of irreducible $G_v = GL(r, F_v)$-modules π_v and (2) equivalence classes of continuous ℓ-adic r-dimensional representations σ_v of $W(\overline{F}_v/F_v)$, namely rank r smooth ℓ-adic sheaves on Spec F_v, with the following properties. It preserves L- and ε-factors of pairs, relates cuspidal π_v with irreducible σ_v, commutes with taking contragredient, and relates the central character of π_v with the determinant of the corresponding σ_v by local class field theory $W(\overline{F}_v/F_v)_{\mathrm{ab}} \simeq F_v^\times$, which is normalized by mapping a geometric Frobenius to a uniformizer in $F_v^\times$. The local correspondence has the property that π and σ correspond if and only if their components π_v and σ_v correspond for all v. Again, Deligne's conjecture was used in the original version of this work as a conjecture, but it is now proven.

To state our fourth main theme, we introduce some notations. Our work relies on Drinfeld's theory of elliptic modules (see [D1, D2]). Their definition and basic properties are discussed in Chaps. 2 and 3. Denote by A the ring of elements of the function field F which are integral outside the fixed place ∞. Let $I \neq \{0\}$ be any ideal in A which is contained in at least two maximal ideals. In Chap. 4 we recall the construction of the (Drinfeld) moduli scheme $X = M_{r,I}$ of isomorphism classes of elliptic A-modules of rank r with I-level structure. It is an affine scheme of finite rank over A. In Chap. 5 we construct a finite étale Galois covering $\widetilde{X}$ of X, whose Galois group Γ is a quotient of an anisotropic inner form $D_\infty^\times$ of $G(F_\infty)$.

Put $\overline{X} = X \otimes_A \overline{F}$, where $\overline{F}$ is a separable closure of F. Let ρ be an irreducible nontrivial representation of Γ and of $D_\infty^\times$. Let π_∞ be the corresponding square-integrable representation of $G(F_\infty)$. In Sect. 6.1 we recall the definition of the associated smooth $\overline{\mathbb{Q}}_\ell$-sheaf $\mathbb{L} = \mathbb{L}(\rho)$ on X and of the $\overline{\mathbb{Q}}_\ell$-adic cohomology spaces $H_c^i(\overline{X}, \mathbb{L})$ of $\overline{X}$ with compact support and coefficients in $\mathbb{L}$. Let $\mathbb{A}_f$ be the ring of F-adèles without a component at ∞, $U = U_I$ the congruence subgroup of $G(\mathbb{A}_f)$ defined by I, and $\mathbb{H}_I$ the Hecke algebra of $\overline{\mathbb{Q}}_\ell$-valued U_I-biinvariant compactly supported functions on $G(\mathbb{A}_f)$. An irreducible $\mathbb{H}_I$-module will be regarded here as an irreducible $G(\mathbb{A}_f)$-module which has a nonzero U_I-fixed vector. The Galois group $\mathrm{Gal}(\overline{F}/F)$ acts on $\overline{F}$, hence on $\overline{X}$ and on $H_c^i = H_c^i(\overline{X}, \mathbb{L}(\rho))$; so does the Hecke algebra $\mathbb{H}_I$. Put $H_c^* = \sum_i (-1)^i H_c^i$; it is a virtual $\mathbb{H}_I \times \mathrm{Gal}(\overline{F}/F)$-module. Namely it is a sum of finitely many irreducibles $\widetilde{\pi}_f \otimes \widetilde{\sigma}$, with integral multiplicities.

Our fourth main theme is the following *explicit reciprocity law*. It underlies the proofs of the purity theorem and the reciprocity law. Suppose that π_∞ is cuspidal. Put $\nu(x) = |x|$ for x in $\mathbb{A}^\times$. Then (1) *each $\widetilde{\pi} = \widetilde{\pi}_f \otimes \pi_\infty$ is cuspidal automorphic and each cuspidal automorphic $G(\mathbb{A})$-module π with component π_∞ can be realized in H_c^* for some $I \neq 0$, (2) the multiplicity of $\widetilde{\pi}_f \otimes \widetilde{\sigma}$ in H^* is one, (3) each $\widetilde{\sigma}$ in H_c^* has $\dim \widetilde{\sigma} = r$, (4) the component $\widetilde{\sigma}_\infty$ is irreducible and corresponds to $\pi_\infty \otimes \nu_\infty^{-(r-1)/2}$ by the local reciprocity law, (5) each ℓ-adic r-dimensional continuous constructible representation σ of $\mathrm{Gal}(\overline{F}/F)$ with irreducible σ_∞ such that $\det(\sigma \otimes \nu^{(r-1)/2}) = 1$ occurs as $\widetilde{\sigma}$ for some I and cuspidal π_∞, and (6) $\widetilde{\pi} \otimes \nu^{-(r-1)/2} \leftrightarrow \widetilde{\sigma}$ is the reciprocity correspondence if $\widetilde{\pi}_f \otimes \widetilde{\sigma}$ occurs in H_c^* for some I.* This explicit law is reduced in Chap. 12 to Deligne's conjecture (= Theorem 6.8).

Moreover we conjecture that (7) $H^i(\overline{X}, \mathbb{L}(\rho)) = H_c^i(\overline{X}, \mathbb{L}(\rho))$ *for all i.* This implies that H^i vanishes unless $i = r - 1$.

The first step in the proof is to decompose the cohomology Hecke×Galois module H_c^* and to show that each irreducible constituent $\pi_f \times \sigma$ has the property that $\pi_f \otimes \pi_\infty$ is cuspidal (automorphic) and each cuspidal representation with the cuspidal component π_∞ occurs, thus establishing the existence of the map $\pi \mapsto \sigma$. This is the heart of the work. It uses the construction of the moduli scheme of elliptic modules and their covering schemes. In the proof of this existence theorem (for each such cuspidal π there exists a corresponding σ) in Chap. 11, we use [in Lemma 11.4(1)] the congruence relations of Theorem 6.10, in addition to the simple trace formula and the Lefschetz fixed point formula.

The second step is to deduce from this existence theorem $\pi \mapsto \sigma$ the local correspondence and its compatibility with the global correspondence. Only in the third and final step, asserting that the map $\pi \mapsto \sigma$ is surjective, is the converse theorem used.

The first major theme in this work, developed in Sect. 6.2, is the congruence relations approach initiated by Eichler–Shimura and Ihara in the case of $GL(2, \mathbb{Q})$. It involves an intrinsic geometric relation concerning Hecke correspondences and powers of the Frobenius. We shall describe here only an application thereof (Theorem 6.10) to the study of eigenvalues. Thus let $\widetilde{\pi}_f \otimes \widetilde{\sigma}$ be an irreducible constituent of H^i as an $\mathbb{H}_I \times \mathrm{Gal}(\overline{F}/F)$-module, for any i ($0 \leq i \leq 2(r-1)$). For almost all v, the component $\widetilde{\pi}_v$ is unramified, and we put $p_{\widetilde{\pi}_v}(t) = \prod_{i=1}^r (t - z_i(\widetilde{\pi}_v))$; the $z_i(\widetilde{\pi}_v)$ are the r Hecke eigenvalues of $\widetilde{\pi}_v$. Also, for almost all v the restriction $\widetilde{\sigma}_v = \widetilde{\sigma}| \mathrm{Gal}(\overline{F}_v/F_v)$ is unramified. Let Fr_v, or $1 \times \mathrm{Fr}_v$, denote any element of $\mathrm{Gal}(\overline{F}/F)$ whose image in $\mathrm{Gal}(\overline{F}_v/F_v)$ is the "arithmetic" Frobenius substitution, and Fr_v^{-1} or $\mathrm{Fr}_v \times 1$ denotes the geometric Frobenius. Then the conjugacy class of $\widetilde{\sigma}(\mathrm{Fr}_v)$ is well defined. Theorem 6.10 asserts that $p_{\widetilde{\pi}_v}(q_v^{-(r-1)/2}\widetilde{\sigma}(\mathrm{Fr}_v^{-1})) = 0$. We conclude that *each eigenvalue u of the geometric Frobenius endomorphism $\widetilde{\sigma}(\mathrm{Fr}_v^{-1})$ is of the form $q_v^{(r-1)/2}z_i(\widetilde{\pi}_v)$ for some Hecke eigenvalue $z_i(\widetilde{\pi}_v)$ (depending on u).* Consequently $\widetilde{\sigma}(\mathrm{Fr}_v^{-1})$ has at most r distinct eigenvalues.

This relates the (geometric) Frobenius and Hecke eigenvalues of $\widetilde{\sigma}$ and $\widetilde{\pi}_f$ which occur together as an irreducible constituent $\widetilde{\pi}_f \otimes \widetilde{\sigma}$ in the composition series of H_c^i, independently of Deligne's conjecture. When combined with the purity theorem, this result asserts that each conjugate of each (geometric) Frobenius eigenvalue (for almost all v) has complex absolute value $q_v^{(r-1)/2}$. This fact is used in the reduction of the existence part (for all π there exists a σ, Theorem 11.1) of the reciprocity law to Deligne's conjecture.

Note that the technique of congruence relations applies with any irreducible $D_\infty^\times$-module ρ, equivalently any square-integrable π_∞. Its applications hold in both cases of cohomology with, and without, compact supports. Also note that the statement $\dim \tilde{\sigma} = r$ for all $\tilde{\sigma}$ can be shown to imply the reciprocity law. In [D1] this is proven in the case of $r = 2$ by means of a different technique.

The work of Part 4 depends on a comparison of the fixed point formula and the trace formula. Since only automorphic $G(\mathbb{A})$-modules occur in the Selberg formula, the purpose of this approach is to show that the $G(\mathbb{A}_f)$-modules $\tilde{\pi}_f$ which occur in the virtual module $H_c^* = \sum_i (-1)^i H_c^i$ are automorphic, in addition to establishing the relation concerning the local Frobenius and Hecke eigenvalues. The Grothendieck fixed point formula gives an expression for the trace of the action of the (geometric) Frobenius $\mathrm{Fr}_v \times 1$ on the cohomology module H_c^* by means of the set of points in $M_{r,I,v}(\overline{\mathbb{F}}_v)$ fixed by the action of the Frobenius and the traces of the resulting morphisms on the stalks of the $\overline{\mathbb{Q}}_\ell$-sheaf $\mathbb{L}(\rho)$ at the fixed points.

In Part 3 we prepare what is needed for this approach. Following [D2], in Chap. 7, the set $M_{r,I,v}(\overline{\mathbb{F}}_v)$ is expressed as a disjoint union of isogeny classes of elliptic modules over $\overline{\mathbb{F}}_v$, and their types are studied. In Chap. 8 it is shown that the elliptic modules with level structure of a given type make a homogeneous space under the action of $G(\mathbb{A}_f)$, and the stabilizer is described. Moreover, the action of the Frobenius Fr_v is identified with multiplication by a certain matrix. A *type* is described in group theoretic terms as an elliptic torus in $G(F)$ (see Definition 7.3), and the cardinality of the set $M_{r,I,v}(\mathbb{F}_{v,n})$ ($[\mathbb{F}_{v,n} : \mathbb{F}_v] = n$) is expressed in terms of orbital integrals of conjugacy classes γ in $G(F)$ which are elliptic in $G(F_\infty)$ and n-admissible (see Definition 8.1) at v.

Next, in Chap. 9, it is shown that the orbital integral at v obtained in Chap. 8 can be expressed as an orbital integral of a spherical function $f_n = f_n^{(r)}$ on G_v whose normalized orbital integral $F(f_n)$ is supported on the n-admissible set. This spherical function is defined by the relation $\mathrm{tr}(\pi_v(z))(f_n) = q_v^{n(r-1)/2} \sum_{i=1}^r z_i^n$. The work of Chap. 9 is independent of the rest of this book. Two different computations of the orbital integrals of f_n are given. The first is representation theoretic. The second (due to a letter of Drinfeld dated March 15, 1976) is elementary.

In Chap. 10 we develop a new form of the trace formula, for a test function $f = \otimes f_w$ with a cuspidal component f_∞ and with component $f_{n,v}^{(r)}$ as above at a place v, where n is sufficiently large compared to the other components f_w ($w \neq v$) of f. It relates the sum of the traces $\mathrm{tr}\,\pi(f)$ of the cuspidal $G(\mathbb{A})$-modules π whose component π_∞ is cuspidal, with a sum of orbital integrals of f at the elliptic γ which are elliptic in $G(F_\infty)$ and n-admissible in $G(F_v)$. The group theoretic side is then the same as that obtained from the stalk side of the fixed point formula, and we are in a position to derive our theorems.

This trace formula is suggested by Deligne's conjecture. For applications of this trace formula in representation theory see [FK2, F3, F4].

Now the scheme $M_{r,I}$ is not proper. The fixed point formula (of Grothendieck) is known in this case only for the powers of the Frobenius (see Theorem 6.6). Hence the components f_w ($w \neq v, \infty$) of the test function f are taken to be the characteristic function of $U_I \cap G_w$. We conclude that each Hecke eigenvalue of the component $\widetilde{\pi}_v$ of $\widetilde{\pi}_f$ which occurs in the virtual module H_c^* is equal to the product by $q_v^{-(r-1)/2}$ of a Frobenius eigenvalue of $\widetilde{\sigma}_v(\mathrm{Fr}_v \times 1)$ for some $\widetilde{\sigma}$ which occurs in the same H_c^*. However, $\widetilde{\pi}_f$ and $\widetilde{\sigma}$ are not shown to appear together as an irreducible constituent $\widetilde{\pi}_f \otimes \widetilde{\sigma}$ of some H_c^i. This establishes the purity theorem, since the Frobenius eigenvalues have complex absolute values of the form $q_v^{c/2}$, where c is an integer, by the Frobenius integrality theorem of Deligne [De3], while the Hecke eigenvalues z_i of the unitarizable cuspidal $\widetilde{\pi}$ satisfy $q_v^{-1/2} < |z_i| < q_v^{1/2}$ for all i.

The same techniques suggest a proof of the reciprocity law as well. Assuming Deligne's conjecture (Theorem 6.8) we have the Hecke algebra, at the places $w \neq v$, at our disposal. The Hecke algebra separates the finitely many $G(\mathbb{A}_f)$-modules which occur in the formulae, and we conclude in Chap. 11 that for each $\widetilde{\pi}_f$ in H^* there is a cuspidal $\pi = \otimes \pi_v$ with $\pi_\infty = \pi_\infty(\rho)$ (cuspidal) and $\widetilde{\pi}_v \simeq \pi_v$ for all v. Moreover, the Hecke eigenvalues of π_v and the Frobenius eigenvalues of $\sigma_v(\mathrm{Fr}_v \times 1)$ are related for almost all v. This implies (using Theorem 6.8) a weak form of the explicit reciprocity law, namely the existence theorem: for each cuspidal π there is a corresponding σ. We also use [in Lemma 11.4(1)] the congruence relations of Theorem 6.10.

In Chap. 12 we reduce the explicit reciprocity law (bijection of π and σ) to its weak form (for each π there is a σ), on using (1) properties of local L and ε-factors, due to Deligne [De2]; (2) the Grothendieck functional equation for ℓ-adic representations σ; (3) Laumon's formula [Lm1], expressing $\varepsilon(\sigma)$ as a product of the local ε-factors of σ; and (4) the simple converse Theorem 13.1 where only cuspidal $\mathrm{GL}(r-1)$-modules with a cuspidal component are used.

As noted above the theory of elliptic modules was introduced by Drinfeld in [D1], [D2] in 1974 and 1977 in order to prove the reciprocity law over function fields, in analogy with the theory of Shimura varieties which had been used by Deligne in 1971 to prove cases of the reciprocity law over number fields. While the theory of Shimura varieties is more amenable to the study of automorphic representations of symplectic, unitary, and orthogonal groups, groups with rather complicated representation theories, Drinfeld's discovery of elliptic modules introduced in the function field case a theory where the prominent group is $\mathrm{GL}(n)$, a group whose representation theory is relatively simple, as it has no nontrivial packets. Drinfeld used his theory to prove the reciprocity law for $\mathrm{GL}(2)$ over a function field, for cuspidal representations with a square-integrable component, and our work is simply an extension to the case of $\mathrm{GL}(r)$, $r \geq 2$. Drinfeld introduced other methods. One is that of Shtukas, developed by Lafforgue [Lf1], [Lf2] in 1997 and 2002, whose aim is to remove the restriction that at least one component be square integrable.

Another, in Am. J. Math. 1983, dealt with unramified representations and initiated Drinfeld's geometric Langlands program.

Deligne lectured in 1975 on Drinfeld's theory of elliptic modules, and the notes appeared in [DH], 1987. Deligne coined the term Drinfeld modules. The idea, in both the number field case of Shimura varieties and the function field case of Drinfeld moduli schemes, is to realize the reciprocity law in suitable cohomology of the moduli scheme, which is a Galois × adèle-bimodule. To determine the constituents one uses a comparison of the automorphic trace formula with the geometric fixed point formula. Since these moduli schemes, Shimura's for elliptic curves and more generally for abelian varieties and Drinfeld's for elliptic modules, are not proper, that is, not compact, the Lefschetz fixed point formula does not apply, and Grothendieck fixed point formula for powers of the Frobenius does not provide sufficient information. Deligne then proposed that the Lefschetz formula holds for a correspondence on a variety over a finite field, provided it is twisted by a sufficiently high—depending on the correspondence—power of the Frobenius.

Of this fundamental conjecture I learned from Kazhdan who proposed holding a seminar at Harvard in 1982 to study Drinfeld's original papers. The present work stems from that seminar and a course given by Kazhdan the following semester. It existed as an unpublished manuscript since 1983, summarized in [FK3] in 1987. The current version was updated during a course I gave at OSU in winter 2012, mainly to add references to literature following [FK3], translate TeX to LaTeX, and restate the reciprocity law in terms of smooth ℓ-adic sheaves in the later part of Chap. 12. Since this work was written at different times, there are some repetitions in it and other expository shortcomings, which I preferred not to eliminate in order to reduce the risk of introducing other lacunae.

Our work concerns cuspidal representations with a cuspidal local component. This permits us to use the simple trace formula, as we developed in purely representation-theoretic context in [FK2, F3, F4, F5], motivated by the present work. That is, much of the complicated trace formula developed by Arthur is not used. More precisely, without Deligne's conjecture, we obtain Ramanujan's conjecture using very little of Arthur's work, and using Deligne's conjecture we obtain all our results on the reciprocity law using only the simple trace formula as in [FK2].

To extend our work from the context of cuspidal representations of $GL(r)$ with a cuspidal component to that of those with a Steinberg component, one needs Arthur's work. This is discussed in [Lm2], 1996, which uses an alternative method of Kottwitz to count the points on the moduli scheme by means of twisted orbital integrals and then proves a "fundamental lemma", relating these to standard orbital integrals. We follow Drinfeld, describing the points directly by orbital integrals.

Another approach was given in [LRS], 1993, to consider instead of $GL(r)$ an inner form thereof. But that meant automorphic representations of the multiplicative group of a division algebra, whose transfer to $GL(r)$ amounts to cuspidal representations with at least two places where the components are square integrable. Here the local reciprocity law could be obtained without using Deligne's conjecture.

To obtain the global reciprocity law, in our work we use Deligne's conjecture and a simple form of the converse theorem, for global representations whose components at a finite nonempty set are cuspidal. In the present draft this simple converse theorem, motivated by our work but of independent interest and techniques far from those of Chaps. 2–12, is delegated to Chap. 13.

Work on Deligne's conjecture was done by Zink [Zi], 1990, who dealt with surfaces, and then by Ed Shpiz in his Harvard thesis [Sp] of 1990 and Pink [P], 1992, both dealing with the case where the variety in question has compactification by a divisor with normal crossings. The Drinfeld moduli scheme is most likely of this type, but this has not been shown as yet. In 1997 appeared Fujiwara's proof [Fu], using rigid analytic techniques, and in 2007 Varshavsky's [V] lucid proof. This made our work hold unconditionally. As noted above L. Lafforgue developed Drinfeld's theory of Shtukas in [Lf1] and [Lf2] of 1997 and 2002 and used [Fu] to prove the reciprocity law (and Ramanujan's conjecture) for all cuspidal representations of $GL(r, \mathbb{A})$, removing the restriction that we put, that at least one component be cuspidal. This important work was used in [DF]. However, in view of the length and depth of [Lf1, Lf2], it seems that there is still merit in our original work, beyond its historical value as the first work where the local and many cases of the global reciprocity law over a function field and in arbitrary rank were proven. Moreover the theory of elliptic, or Drinfeld, modules is of interest in its own right, as evidenced from the wealth of literature on it, and so is our simple converse Theorem 13.1.

Within each chapter, theorems, propositions, and lemmas are numbered consecutively, and so are—separately—definitions, remarks, and examples.

I wish to express my very deep gratitude to David Kazhdan for teaching me the theory of Drinfeld modules, first in a seminar which we had on Drinfeld's papers [D1, D2], then in a course he gave, and also in numerous conversations relating to these notes. The results and some of the techniques of this work are exposed in our joint note [FK3].

I wish to thank the referees for useful comments.

Recent support by the Humboldt Stiftung at Berlin's HU, MPIM-Bonn, SFB at Bielefeld, Lady Davis Foundation at the Hebrew University, Newton Institute at Cambridge, IHES, and the Fulbright Foundation is warmly acknowledged.

Part 1. Elliptic Moduli

This part consists of an exposition to Drinfeld's theory of elliptic modules (see [D1, D2]). Their definition and basic properties are discussed in Chaps. 2 and 3. Denote by A the ring of elements of the function field F which are integral outside the fixed place ∞. Let $I \neq \{0\}$ be any ideal in A which is contained in at least two maximal ideals. In Chap. 4 we recall the construction of the (Drinfeld) moduli scheme $X = M_{r,I}$ of isomorphism classes of elliptic A-modules of rank r with I-level structure. It is an affine scheme of finite rank over A. In Chap. 5 we construct a finite étale Galois covering $\widetilde{X}$ of X, whose Galois group Γ is a quotient of an anisotropic inner form $D_\infty^\times$ of $G(F_\infty)$.

2. ELLIPTIC MODULES: ANALYTIC DEFINITION

Let p be a prime number, d a positive integer, $q = p^d$, $\mathbb{F}_q$ a field of q elements, C an absolutely irreducible smooth projective curve defined over $\mathbb{F}_q$, and F the function field $\mathbb{F}_q(C)$ of C over $\mathbb{F}_q$, that is, the field of rational functions on C over $\mathbb{F}_q$. At each place v of F, namely a closed point of C, let F_v be the completion of F at v and A_v the ring of integers in F_v. Fix a place ∞ of F. Let C_∞ be the completion of an algebraic closure $\overline{F}_\infty$ of F_∞.

Let $A = H^0(C - \{\infty\}, \mathcal{O}_C)$ be the ring of regular functions on $C - \{\infty\}$, namely the ring of functions in F whose only possible poles are at ∞. For each v in $\operatorname{Spec} A = C - \{\infty\}$, the quotient field $\mathbb{F}_v = A/v$ is finite. Denote its cardinality by q_v. Note that A_v is the completion of A at v. For any a in A let $(a) = aA$ be the ideal in A generated by a. Let $|.| = |.|_\infty$ be the absolute value on A which assigns to $a \neq 0$ in A the cardinality of the quotient ring $A/(a)$. It extends uniquely to $F, F_\infty, \overline{F}_\infty$, and C_∞. Let π_∞ be a generator of the maximal ideal of A_∞. Let q_∞ be the cardinality $|A_\infty/\pi_\infty|$ of the finite field A_∞/π_∞. If $a \in A$ has a pole of order n at ∞, then $|a|_\infty = |\pi_\infty^{-n}| = q_\infty^n$.

A function f from C_∞ to C_∞ is called *entire* if it is equal to an everywhere convergent power series. Thus $f = \sum_0^\infty a_n x^n$ $(a_n \in C_\infty)$, where $|a_n|^{1/n} \to 0$.

Lemma 2.1. *Let f be a nonconstant entire function on C_∞. Then f attains each value of C_∞.*

Proof. This is the same as the proof in the case of characteristic zero. See [Ko], Ex. 13, Section IV.4 (p. 108), where the lemma is proven with C_∞ replaced by the completion Ω of the algebraic closure $\overline{\mathbb{Q}}_p$ of $\mathbb{Q}_p$. $\square$

A quotient $f = h/g$ of two entire functions h, g on C_∞, with $g \not\equiv 0$, is called a *meromorphic function* on C_∞. The *divisor* $\operatorname{Div} f$ of a meromorphic function f on C_∞, with zeroes a_i and poles b_j of multiplicities n_i and m_j (respectively), is the formal sum $\sum_i n_i(a_i) - \sum_j m_j(b_j)$.

Corollary 2.2. *Let f, g be entire functions on C_∞ with $\operatorname{Div} f = \operatorname{Div} g$. Then there is $c \neq 0$ in C_∞ with $f = cg$.*

Proof. If $g \not\equiv 0$ then f/g is entire, as its Taylor expansion at 0 converges everywhere. But f/g has no zeroes. Hence it is constant by Lemma 2.1. $\square$

A set L in C_∞ is called *discrete* if for each positive number c the set $\{x$ in $L; |x| \leq c\}$ is finite. Since C_∞ is a non-Archimedean field, then for each discrete set L there is an entire function e_L with $\operatorname{Div} e_L = L$. If L contains zero then there is a unique e_L normalized so that $e'_L(0) = 1$. It is given by the product

$$e_L(x) = x \prod_a (1 - x/a) \qquad (a \neq 0 \text{ in } L).$$

Proposition 2.3. *Let L be an additive discrete subgroup of C_∞. Then e_L defines an isomorphism from C_∞/L to C_∞ as additive groups.*

Y.Z. Flicker, *Drinfeld Moduli Schemes and Automorphic Forms: The Theory of Elliptic Modules with Applications*, SpringerBriefs in Mathematics, DOI 10.1007/978-1-4614-5888-3_2, © Yuval Z. Flicker 2013

Proof. (i) From Lemma 2.1 it follows that e_L defines a set theoretic surjection from C_∞/L to C_∞. (ii) To show that e_L is a group homomorphism, we first consider the case where L is finite. It is clear from the definition of e_L that $e_L(x+y)-e_L(x)-e_L(y) = 0$ if x or y lie in L. Hence the polynomial $e_L(x)e_L(y)$, whose degree in x is $|L|$, divides the polynomial $e_L(x+y)-e_L(x)-e_L(y)$, whose degree in x is less than $|L|$. We conclude that $e_L(x + y) = e_L(x) + e_L(y)$. In general we can write L as a union of finite subgroups L_n. Then $e_L = \lim_n e_{L_n}$, and (ii) follows. (iii) Since the kernel of e_L is L, the proposition follows from (i) and (ii). □

Definition 2.1. A *lattice* L is a discrete, finitely generated A-submodule of C_∞.

Lemma 2.4. *A finitely generated module over a Dedekind domain (an integral domain in which every nonzero proper ideal factors into a product of prime ideals) is projective if and only if it is torsion free.*

Proof. See [BN], VII, Section 4.10, Prop. 22 (p. 543). □

Since C_∞ is a field, the lattice L is a torsion-free A-module. Since A is a Dedekind domain, L is projective. Denote by $r = \mathrm{rank} L$ the rank of the lattice L. We have

Lemma 2.5. *For any $a \neq 0$ in A there is an isomorphism from L/aL to $(A/aA)^r$.*

The isomorphism $e_L : C_\infty/L \to C_\infty$ and the A-module structure on C_∞/L define an A-module structure on C_∞ by $ax = \varphi_{a,L}(x) = e_L\left(a\left(e_L^{-1}(x)\right)\right)$ (a in A, x in C_∞).

Lemma 2.6. *For each a in A the function $\varphi_{a,L}$ is a polynomial of degree $|a|^r$ over C_∞.*

Proof. Put $\psi_{a,L}(x) = e_L(ax)$. The kernel of $\psi_{a,L}$ is $a^{-1}L$. Hence there is some $c \neq 0$ with $\psi_{a,L}(x) = c\prod_b (e_L(x) - e_L(b))$ (b in $a^{-1}L/L$). Consequently $\varphi_{a,L}(x) = c\prod_b (x - e_L(b))$ is a polynomial over C_∞ whose degree is equal to the cardinality $|a|^r$ of L/aL. □

Let E_∞ be a fixed finite extension of F_∞ in C_∞. Let $\overline{E}_s$ denote the completion of the separable closure E_s of E_∞ in C_∞. The fields E_∞, E_s, and $\overline{E}_s$ appear only in Chap. 2.

Definition 2.2. A lattice L is called a *lattice over* E_∞ if it lies in E_s and it is invariant under the action of the Galois group $\mathrm{Gal}(E_s/E_\infty)$ of E_s over E_∞.

Example 2.1. The ring $L = A$ is a lattice over F_∞, of rank one.

Proposition 2.7. *If L is a lattice over E_∞ then $\varphi_{a,L}$ is a polynomial over E_∞.*

Proof. The coefficients of the Taylor expansion at 0 of e_L lie in $\overline{E}_s$, and they are invariant under $\mathrm{Gal}(E_s/E_\infty)$ by definition of L. Hence they lie in E_∞ (by [Ta], Theorem 1, p. 176). The proposition now follows from the proof of Lemma 2.6. □

Definition 2.3. (i) The lattices L, L' over E_∞ are *isomorphic* if $L' = uL$ for some $u \neq 0$ in E_∞. (ii) Let L, L' be lattices of rank r. A *morphism* from L to L' is u in E_∞ with $uL \subset L'$.

Lemma 2.8. *If L is a lattice over E_∞ and $u \neq 0$ is in E_∞, then $u^{-1}\varphi_{a,uL}(ux) = \varphi_{a,L}(x)$.*

Proof. Using the identity $e_{uL}(x) = ue_L(u^{-1}x)$ we rewrite the relation $\varphi_{a,uL}$ $(e_{uL}(x)) = e_{uL}(ax)$ in the form $\varphi_{a,uL}(ue_L(u^{-1}x)) = ue_L(au^{-1}x)$. This implies the required identity

$$u^{-1}\varphi_{a,uL}(ue_L(x)) = e_L(ax) = \varphi_{a,L}(e_L(x)).$$

$\square$

Definition 2.4. A polynomial h in $C_\infty[x]$ is called *additive* if $h(x + y) = h(x) + h(y)$.

Lemma 2.9. *If h is additive then $h(x) = \sum_{i=1}^{I} a_i x^{p^i}$.*

Proof. If $h(x) = \sum b_i x^i$ is additive, then $b_i((x + y)^i - x^i - y^i) = 0$. If $i = p^n j$ with $j > 1$ prime to p, then $(x + y)^i = \left(x^{p^n} + y^{p^n}\right)^j$ is not equal to $x^i + y^i$, since it has the term $jx^{p^n(j-1)}y^{p^n}$ in its binomial expansion. Hence $b_i = 0$ if i is not a power of p. $\square$

The map $\varphi_L : a \mapsto \varphi_{a,L}$ has several properties which suggest the following:

Definition 2.5. (1) A map $\varphi : A \to E_\infty[x]$, $a \mapsto \varphi_a$, is called an *elliptic module* of rank r over E_∞ if (i) $\varphi_a(x + y) = \varphi_a(x) + \varphi_a(y)$ (a in A); (ii) $\varphi_{ab} = \varphi_a \circ \varphi_b$, $\varphi_{a+b} = \varphi_a + \varphi_b$; (iii) $\deg \varphi_a = |a|^r$; and (iv) $\varphi_a(x) \equiv ax \bmod x^p$.

  (2) The elliptic modules φ, φ' are *isomorphic* if there is $u \neq 0$ in E_∞ with $\varphi'_a(x) = u\varphi_a(u^{-1}x)$ (a in A).

  (3) Let φ, φ' be two elliptic modules of rank r over E_∞. A *morphism* from φ to φ' is an additive polynomial P in $E_\infty[x]$ with $P \circ \varphi_a = \varphi'_a \circ P$ (a in A).

Lemma 2.10. *Any morphism P is of the form $P(x) = \sum_i a_i x^{q^i}$, where a_i lie in E_∞. The group of automorphisms of an elliptic module is $\mathbb{F}_q^\times$.*

Proof. For any a in the finite subfield $\mathbb{F}_q$ of A we have $\varphi_a(x) = ax$ and $\varphi'_a(x) = ax$. Hence $aP(x) = P(ax)$, and the lemma follows. $\square$

Corollary 2.11. (1) *For each b in A, we have $\varphi_b(x) = \sum_{i=1}^{I(b)} a_i x^{q^i}$, where $I(b) = rv_q(b)$, $v_q(b) = \log_q |b|$, and $a_{I(b)} \neq 0$. (2) If $A = \mathbb{F}_q[t]$ then $|t| = q$, and an elliptic module is determined by $\varphi_t(x) = tx + \sum_{i=1}^{r} a_i x^{q^i}$ with $a_r \neq 0$. (3) In (2), up to isomorphism we may replace a_i by $a_i u^{q^i - 1}$.*

Remark 2.1. (1) An elliptic module of rank r over C_∞ is defined on replacing E_∞ by C_∞ in Definition 2.5(1). The following theorem holds also with E_∞ replaced by C_∞. (2) Since the case of $r = 0$ is trivial, we consider from now on only the case of $r > 0$.

Theorem 2.12. *The map $L \mapsto \varphi_L$ defines an equivalence from the category of (isomorphism classes of) lattices of rank r over E_∞ to the category of (isomorphism classes of) elliptic modules of rank r over E_∞.*

Proof. (i) Our first aim, accomplished in (iv), is to construct an inverse to the map $L \mapsto \varphi_L$. Thus let φ be an elliptic module over E_∞, of rank r. Fix a in $A - \mathbb{F}_q$; then $|a| > 1$. We have $\varphi_a(x) = ax + \sum_i a_i x^{q^i}$ with a_i in E_∞ ($1 \le i \le s = rv_q(a)$). We claim that there exists a unique power series $e(x) = \sum_{i=0}^{\infty} e_i x^{q^i}$ with $e_0 = 1$, e_i in E_∞, and $\varphi_a(e(x)) = e(ax)$. To show this we equate the coefficients of x^{q^n} in $\varphi_a(e(a^{-1}x)) = e(x)$ to obtain

$$e_n\left(1 - a^{1-q^n}\right) = a_n a^{-q^n} + \sum_{i=1}^{n-1} a_i e_{n-i}^{q^i} a^{-q^n}.$$

($a_n = 0$ for $n > s$; $e_i = 0$ for $i < 0$); this yields a recursive formula for e_n.

(ii) We claim that e is entire. To see this we note that for $n > s$ we have

$$e_n(a^{q^n} - a) = \sum_{i=1}^{s} a_i e_{n-i}^{q^i}.$$

Then

$$|a| r_n \le \max\left\{|a_i|^{p^{-n}} r_{n-i}; 1 \le i \le s\right\},$$

where $r_j = |e_j|^{q^{-j}}$. For θ with $|a|^{-1} < \theta < 1$, there is n' such that for $n > n'$ we have $r_n \le \theta \max\{r_{n-i}; 1 \le i \le s\}$. Hence $r_n \to 0$, and e is entire.

(iii) For any b in A we claim that $\varphi_b(e(x)) = e(bx)$. Indeed, if $b \neq 0$, then we have

$$(\varphi_b \circ e \circ b^{-1})(x) = (\varphi_b \circ \varphi_a \circ e \circ a^{-1} \circ b^{-1})(x) = (\varphi_a \circ (\varphi_b \circ e \circ b^{-1}) \circ a^{-1})(x).$$

But then the uniqueness of the solution e for the equation $\varphi_a \circ e \circ a^{-1} = e$ implies the claim.

(iv) Let L be the kernel of e. Since the derivative e' of e is identically one, the zeroes of e are simple. Hence L lies in E_s. The group L is a discrete, $\mathrm{Gal}(E_s/E_\infty)$-invariant A-module, and we have $|L/aL| = |a|^r$. Hence L is finitely generated. Indeed, if $\{b_i\}$ are $|a|^r$ representatives in L for L/aL, then the finite set of x in L with $|x| \le \max_i\{|b_i|\}$ generates L as an A-module. Now since L is torsion free and A is a Dedekind domain, L is flat. A finitely generated flat module over a Noetherian ring is projective. Hence L is a lattice of rank r. Since we have $e = e_L$ and $\varphi_{a,L} = \varphi_a$ for all a in A, we constructed an inverse to the map $L \mapsto \varphi_L$, establishing a set theoretic isomorphism.

(v) Let L, L' be lattices of rank r over E_∞ with $uL \subset L'$ for some u in E_∞. Then $e_{L'}(ux)$ is L-invariant. The proof of Lemma 2.6 shows that there is a polynomial P over E_∞ with $P(e_L(x)) = e_{L'}(ux)$. But then P is additive, and

$$(P \circ \varphi_{a,L})(e_L(x)) = P(e_L(ax)) = e_{L'}(uax)$$
$$= \varphi_{a,L'}((e_{L'} \circ u)(x)) = (\varphi_{a,L'} \circ P)(e_L(x))$$

implies that P is a morphism from φ_L to $\varphi_{L'}$.

(vi) Conversely, if P is a polynomial over E_∞ with $P \circ \varphi_L = \varphi_{L'} \circ P$, then

$$(P \circ e_L)(x) = (P \circ \varphi_{a,L} \circ e_L)(a^{-1}x) = (\varphi_{a,L'}(P \circ e_L))(a^{-1}x).$$

Hence we conclude from the uniqueness assertion of (i) that there is $u \neq 0$ in E_∞ with $(P \circ e_L)(x) = e_{L'}(ux)$. Then $uL \subset L'$, and the theorem follows.

$\square$

Remark 2.2. It is clear from the proof of (vi) that any polynomial P in $E_\infty[x]$ with $P \circ \varphi_a = \varphi'_a \circ P$ for all a in A has to be additive.

3. Elliptic Modules: Algebraic Definition

Definition 2.5 of an elliptic module over a field extension of F_∞ is purely algebraic. So it has a natural generalization defining elliptic modules over any field over A.

Let B be a ring of characteristic p (by definition, $p = 0$ in B). Let R be a B-algebra, namely a ring with a ring homomorphism $i : B \to R$. Then R has characteristic p. By an *affine group scheme* G over B we mean a representable functor from the category of B-algebras R to the category of groups; thus $G(R) = \mathrm{Hom}(B', R) = \mathrm{Hom}(\mathrm{Spec}\,R, \mathrm{Spec}\,B')$ if G is represented by the B-algebra B'. Then B' is a Hopf algebra (with comultiplication $\mu : B' \to B' \otimes_B B'$, counite ε, and coinverse ι, which are B-algebra homomorphisms satisfying standard axioms), and $\mathrm{Spec}\,B'$ has a natural structure of a group. As usual, we write $G = \mathrm{Spec}\,B'$ if the functor G is representable by B'. The group G is called *algebraic* if B' is finitely generated over B. For example, the *additive group* $\mathbb{G}_{a,B}$ over B is the functor which associates to the B-algebra R the additive group of R. Then $\mathbb{G}_{a,B}(R) = \mathrm{Hom}_B(B[x], R)$, namely $\mathbb{G}_{a,B}$ is represented by the B-algebra $B[x]$ (which is of finite type), and $\mathbb{G}_{a,B} = \mathrm{Spec}\,B[x]$. The group structure on $\mathbb{G}_{a,B}$ is defined by the Hopf algebra structure on $B[x]$, namely by the comultiplication $\mu : B[x] \to B[x] \otimes_B B[x]$, $x \mapsto x \otimes 1 + 1 \otimes x$, counit $\varepsilon : x \to 0$, and coinverse $\iota : x \to -x$.

The set of morphisms from a functor E to a functor E' is denoted by $\mathrm{Hom}(E, E')$. When $E = E'$ we write $\mathrm{End}\,E$ for $\mathrm{Hom}(E, E)$. Put $\mathrm{End}_B\,\mathbb{G}_a$ for $\mathrm{End}\,\mathbb{G}_{a,B}$. If $G' = \mathrm{Spec}\,B'$, $G'' = \mathrm{Spec}\,B''$ are affine group schemes then the map which associates to the Hopf algebra morphism $P : B'' \to B'$ the morphism $P : G' \to G''$, defined by $P(u : B' \to R) = (u \circ P : B'' \to R)$, is an isomorphism $\mathrm{Hom}(B'', B') \widetilde{\to} \mathrm{Hom}(G', G'')$. Let $B[\tau]$ be the ring generated by τ over B under the relations $\tau b = b^p \tau$ for all b in B.

Lemma 3.1. *The ring* $\mathrm{End}_B\,\mathbb{G}_a$ *is canonically isomorphic to* $B[\tau]$ *as a ring.*

Proof. An endomorphism P of $\mathbb{G}_{a,B}$ is equivalent to a homomorphism $B[x] \to B[x]$ over B which commutes with μ. Such P is determined by the image $P(x) = \sum_i a_i x^i$ of x. The morphism P commutes with μ if and only if $P(y + z) = P(y) + P(z)$, where $y = x \otimes 1$ and $z = 1 \otimes x$. The proof of Lemma 2.9 implies that $P(x) = \sum b_i x^{p^i}$. Denote by τ the endomorphism $\tau(x) = x^p$. For b in B, denote by b also the endomorphism $b(x) = bx$ of multiplication by b. Then $\tau b = b^p \tau$. Hence $P = \sum b_i \tau^i$ lies in $B[\tau]$. Since $B[\tau]$ clearly lies in $\mathrm{End}_B\,\mathbb{G}_a$, the lemma follows. $\square$

Remark 3.1. (1) We identify B with its image in $B[\tau]$. (2) If the characteristic of B is zero then $\mathrm{End}_B\,\mathbb{G}_a = B$: the only endomorphisms of $\mathbb{G}_a$ over B are $b(x) = bx$. The richer structure of $\mathrm{End}_B\,\mathbb{G}_a$ in the case of characteristic $p > 0$ is the basis of the theory of elliptic modules.

Definition 3.1. (i) Let $D\left(\sum_0^n b_i \tau^i\right) = b_0$ define the ring homomorphism $D : B[\tau] \to B$. (ii) The polynomial P in $B[\tau]$ is called *separable* if $D(P) \neq 0$.

Y.Z. Flicker, *Drinfeld Moduli Schemes and Automorphic Forms: The Theory of Elliptic Modules with Applications*, SpringerBriefs in Mathematics, DOI 10.1007/978-1-4614-5888-3_3, © Yuval Z. Flicker 2013

We can now make the following:

Definition 3.2. Let K be a field over A. An *elliptic module* over K is a ring homomorphism $\varphi : A \to K[\tau] = \operatorname{End}_K \mathbb{G}_a$, $a \mapsto \varphi_a$, whose image is not contained in K, such that $D \circ \varphi = i$.

Lemma 3.2. *Any elliptic module* $\varphi : A \to K[\tau]$ *is an embedding.*

Proof. Since $K[\tau]$ has no zero divisors, the ideal $\ker \varphi$ is prime. Suppose that $\ker \varphi \neq 0$. Since A is a Dedekind domain, $\ker \varphi$ is a maximal ideal. Hence the image of φ is a field, necessarily contained in K, since all invertible elements in $K[\tau]$ lie in K. We obtain a contradiction to the definition of an elliptic module. This implies that $\ker \varphi = 0$, as required. $\qquad\square$

To compare Definition 3.2 with Definition 2.5, define the *degree* homomorphism $\deg : K[\tau] \to \mathbb{Z}$ by $\deg \left(\sum_0^n b_i \tau^i \right) = p^n$ if $b_n \neq 0$, and $\deg 0 = 0$. It satisfies

$$\deg(x + y) \leq \max(\deg x, \deg y), \qquad \deg(xy) = \deg x \cdot \deg y.$$

Hence the homomorphism $d : A \to \mathbb{Z}$, $d(a) = \deg \varphi_a$, satisfies $d(ab) = d(a)d(b)$ and $d(a + b) \leq \max(d(a), d(b))$. Lemma 3.2 then implies that d is an absolute value on A at some point of C, in fact at ∞ since $d(a) \geq 1$ on A. Hence there exists some $r > 0$, which we call the *rank* of φ and denote by $\operatorname{rank} \varphi$, with $d(a) = |a|^r$. Once we show, in Corollary 3.5, that r is an integer, it is clear that when K is the extension E_∞ of F_∞ as in Definition 2.5, the elliptic module of Definition 3.2 is an elliptic module in the sense of Definition 2.5, with rank r.

Definition 3.3. (*i*) The *characteristic* of an elliptic module $\varphi : A \to K[\tau]$ is the prime ideal $\ker[i : A \to K]$ in $\operatorname{Spec} A$. It is denoted by $\operatorname{char} \varphi$. (*ii*) Let E_φ be the functor from the category of K-algebras R to the category of A-modules, whose composition with the forgetful functor from A-modules to groups is $\mathbb{G}_{a,K}$, with the A-structure defined by φ. Thus if $\varphi_a = i(a) + \sum_{i=1}^S b_i \tau^i$ (b_i in K), then $a \circ r = \varphi_a(r) = i(a)r + \sum b_i r^{p^i}$ in $E_\varphi(R)$ (r in R). A functor E is called an *elliptic functor* if it is isomorphic to E_φ, where φ is an elliptic module. (*iii*) For any ideal I in A, let $E_I = \operatorname{Ann}(I) | E$ be the subfunctor of the elliptic functor E annihilated by I. Thus if $E = E_\varphi$ then $E_I(R)$ is the A/I-module consisting of all x in $E(R)$ with $a \circ x = 0$ for all a in I.

Remark 3.2. (1) Since A is a Dedekind domain, either $\operatorname{char} \varphi = \{0\}$, in which case $i : A \to K$ is injective, or $\operatorname{char} \varphi$ is a maximal ideal. (2) If $I = (a)$ is principal and $E = E_\varphi$, write E_a for $E_I = \ker \varphi_a$. In general, $E_I(R) = \cap E_a(R)$ (a in I). If $a_1, \ldots, a_s$ generate I in A then $E_I(R) = E_{a_1}(R) \cap \cdots \cap E_{a_s}(R)$. (3) The functors E and E_a are equal if and only if $a = 0$, and $E_I = E$ if and only if $I = \{0\}$.

Lemma 3.3. *Let w be a maximal ideal of A. Let π denote a uniformizer in the local ring A_w. Let E'' be an $A_w/\pi^{2m} A_w$-module of cardinality $|\pi|^{-2mr}$. Suppose the submodule $E' = \ker \pi^m \mid E''$ has cardinality $|\pi|^{-mr}$. Then E' is a free $A_w/\pi^m A_w$-module of rank r.*

Proof. As a finitely generated torsion module over a local ring A_w, we have that E'' is a finite direct sum of modules isomorphic to $A_w/\pi^j A_w$. Since π^{2m} annihilates E'' we have that $0 < j \leq 2m$. Now $\ker \pi^m$ is the sum of $A_w/\pi^{\min\{j,m\}} A_w$. Hence $|\ker \pi^m|^2 \geq |E''|$, with equality only when $j = 2m$ for all j. The lemma follows. $\square$

Let φ be an elliptic module of rank r over K. Fix an algebraic closure $\overline{K}$ of K.

Theorem 3.4. *Let I be an ideal in A which is prime to char φ. Then $E_I(\overline{K})$ is a free A/I-module of rank r.*

Corollary 3.5. (i) *Fix w in $\operatorname{Spec} A$ prime to char φ, and b in $A - \mathbb{F}_q$ whose only zero is at w. Then $\varinjlim_m E_{b^m}(\overline{K})$ is a free F_w/A_w-module of rank r.* (ii) *The rank of an elliptic module is an integer.*

Remark 3.3. The ideal I is prime to char φ if and only if I is not contained in char φ.

Proof.   (i) Let w in $\operatorname{Spec} A$ be prime to char φ. We first consider the case where $I \neq A$ is a principal ideal generated by b in A whose only zeroes are at w. Then (b) is prime to $\ker i = \operatorname{char} \varphi$, so that $i(b) \neq 0$. We have $F_b(\overline{K}) = \ker \varphi_b$, where $\varphi_b(x) = i(b)x + \cdots + a_s x^{|b|^r}$, and $a_s \neq 0$. Since $d\varphi_b/dx (\equiv i(b))$ is never zero, the roots of $\varphi_b(x) = 0$ are simple, and the cardinality of $E_b(\overline{K})$ is $|b|^r$. The same argument implies that the cardinality of $E_{b^2}(\overline{K})$ is $|b|^{2r}$. In the notations of Lemma 3.3 there is some positive integer m with $bA_w = \pi^m A_w$, and we have $A/bA \simeq A_w/\pi^m A_w$ and $A/b^2 A \simeq A_w/\pi^{2m} A_w$. The theorem follows on taking $E'' = E_{b^2}(\overline{K})$ and $E' = E_b(\overline{K})$ in Lemma 3.3, in our case where $I = (b)$ is principal and supported at w. In particular the corollary is proven.

(ii) Let I be any ideal which is prime to char φ. Since A is Noetherian I has the primary decomposition $I = \cap w^{n(w)}$, where w are maximal ideals in $\operatorname{Spec} A$ not contained in char φ, and $n(w)$ are positive integers. Then $A/I = \oplus A/w^{n(w)}$, and any finitely generated A/I-module M is the direct sum of its primary components M_w, which are $A/w^{n(w)}$-modules. To show that $E_I(\overline{K})$ is a free A/I-module of rank r it suffices to show that $E_{I^h}(\overline{K})$ is a free A/I^h-module of rank r for some positive integer h. We take h to be the (finite) cardinality of the quotient of the multiplicative group of ideals in A by its subgroup of principal ideals. Then $w^{h \, n(w)}$ is principal, and the theorem follows from the case proven in (i).

$\square$

Let w be a prime ideal of A which is not contained in char φ. Let b be an element of $A - \mathbb{F}_q$ whose only zero is at w. In view of Corollary 3.5 we can make

Definition 3.4. The *Tate module* $T_w E$ of E at w is $\mathrm{Hom}_{A_w}(F_w/A_w,$ $\varprojlim_m E_{b^m}(\overline{K}))$.

Corollary 3.6. *If w is prime to $\mathrm{char}\,\varphi$ then $T_w E$ is a free A_w-module of rank r.*

Let K be a ring over A. Let φ, φ' be two elliptic modules over K. Let $E = E_\varphi$ and $E' = E_{\varphi'}$ be the associated elliptic functors. In the rest of this chapter we study basic properties of the group $\mathrm{Hom}(E, E')$ which are fundamental for the description in Part 2 of isogeny classes.

Lemma 3.7. *The group $\mathrm{Hom}(E, E')$ is isomorphic to the group of all P in $K[\tau]$ with $P\varphi_a = \varphi'_a P$ for all a in A.*

Proof. This follows at once from the definitions. Namely, since the composition of E or E' with the forgetful functor from A-modules to groups is $\mathbb{G}_{a,K}$, any P in $\mathrm{Hom}(E, E')$ lies in $\mathrm{End}_K \mathbb{G}_a = K[\tau]$. The morphism P commutes with the A-module structures on E and E' if and only if $P\varphi_a = \varphi'_a P$ for all a in A. $\square$

Remark 3.4. The group $\mathrm{Hom}(E, E')$ has an A-module structure given by $a \circ P = \varphi'_a P$.

Lemma 3.8. *The A-module $\mathrm{Hom}(E, E')$ is torsion free.*

Proof. There are no zero divisors in $K[\tau]$. $\square$

Let $G' = \mathrm{Spec}\, B'$ and $G = \mathrm{Spec}\, B$ be affine group schemes over K. We say that the morphism $G \to G'$ is *injective* (resp. *surjective*) if the corresponding morphism $B' \to B$ of Hopf K-algebras is surjective (resp. injective). By a *subgroup* (or group subscheme) of G we mean a pair (H, i) consisting of a group H and an injection $i : H \to G$. A subgroup H of G corresponds to a Hopf ideal I in B, namely an ideal I with $\mu(I) \subseteq I \otimes B + B \otimes I$, $\varepsilon(I) = 0$, and $\iota(I) \subseteq I$.

The subgroup H of G is called *normal* if $H(R)$ is normal in $G(R)$ for all K-algebras R.

The *kernel* $H = H_P$ of a morphism $P : G \to G'$ is the functor defined by $H(R) = \ker[P(R) : G(R) \to G'(R)]$. The kernel H is the (normal) subgroup of G corresponding to the ideal $I_P = \widetilde{P}(\ker \varepsilon')B$ of B, where $\varepsilon' : B' \to K$ is the counit of B' and $\widetilde{P} : B' \to B$ is the Hopf algebra morphism corresponding to P. Indeed, $g : B \to R$ in $G(R)$ is in the kernel $H(R)$ if and only if $g \circ \widetilde{P} : B' \to R$ is zero in $G'(R)$, namely $g \circ \widetilde{P}$ factorizes through ε', or $g(\widetilde{P}(\ker \varepsilon')) = 0$, equivalently $g(I_P) = 0$.

A *quotient* of G is a pair (G', P) consisting of an affine group scheme $G' = \mathrm{Spec}\, B'$ and a surjection $P : G \to G'$. If H is a normal subgroup of G then there is a unique quotient $P : G \to G'$ with kernel H (see [W], (16.3)). The quotient (G', P) of G has the universal property that for any morphism $Q : G \to G''$ which vanishes on the kernel H_P of $P : G \to G'$, namely Q satisfies $I_Q \subset I_P$, there is a unique morphism $R : G' \to G''$ with $Q = RP$ (see [W], (15.4)).

The group G is called a *finite* group scheme of *order* m over K if B is an algebra of rank m over K. A finite group G over K is called *étale* if $B \otimes_K K_s$ is a finite direct product of copies of the separable closure K_s of K in $\overline{K}$.

An affine group G is called *connected* if Spec B is connected (equivalently, irreducible). In the case where G is the additive group $\mathbb{G}_{a,K}$, all subgroups and quotients can be explicitly described as follows.

Lemma 3.9. (i) *Every proper subgroup H of $\mathbb{G}_{a,K}$ is of the form H_P for some $P \neq 0$ in $\mathrm{End}_K \mathbb{G}_a$. Every quotient of $\mathbb{G}_{a,K}$ is of the form $P : \mathbb{G}_{a,K} \to \mathbb{G}_{a,K}$. The quotient map corresponding to H_P is defined by $P : K[x] \to K[x]$, $x \mapsto P(x)$. We have $H_P = H_Q$ if and only if $P = aQ$ for some $a \neq 0$ in K. The order of H_P is the degree of $P(x)$. (ii) The morphism P is separable if and only if H_P is étale. The group H_P is connected if and only if $P = a\tau^j$ for some $a \neq 0$ in K and $j \geq 0$.*

Proof.   (i) If H is a proper subgroup of $\mathbb{G}_{a,K}$ then it corresponds to a proper ideal I in the principal ideal domain $K[x]$. Hence H is finite, and $I = (P(x))$ for some monic $P(x)$ in $K[x]$. The ideal I is an Hopf ideal precisely when $\varepsilon(P) = 0$, thus $P(0) = 0$, and $\mu(P(x)) = P(x+y)$ lies in $I \otimes K[x] + K[x] \otimes I$. Namely there are polynomials a_i and b_j in $K[x]$ $(0 \leq i \leq n, 0 \leq j \leq m)$ with $a_n \neq 0$, $b_m \neq 0$, and

$$P(x+y) = P(x) \sum_{0 \leq i \leq n} a_i(y)x^i + P(y) \sum_{0 \leq j \leq m} b_j(y)x^j.$$

Choose $n \geq 0$ to be minimal. Comparing highest powers of x we conclude that $m = n + \deg(P)$ and $a_n(y) + P(y)b_m(y) = 0$, namely n can be reduced by 1 if $n > 0$. Thus $n = 0$. By symmetry (of x and y), the $b_i(y)$ are independent of y. Thus $P(x+y) = P(x)a_0(y) + P(y)c_0(x)$. Comparing the highest degree terms of $P(x+y)$, $P(x)a_0(y)$, and $P(y)c_0(x)$ we see that $c_0(x) = 1$ and $a_0(y) = 1$. Hence P is additive, and (i) follows.

(ii) The group of connected components of the finite group H is isomorphic to the group $H(\overline{K})$ of points on H; this is isomorphic to the group of zeroes of $P(x) = 0$. The group H is connected if and only if $P(x)$ has only one zero, necessarily at $x = 0$. Since P is additive, it has the asserted form $P(x) = ax^{p^j}$; (ii) follows.

$\square$

We shall now verify directly the universal property of quotients in the case of $\mathbb{G}_{a,K}$.

Lemma 3.10. *Let $P(x)$, $\Phi(x)$ be additive polynomials in $K[x]$ such that $P(x)$ divides $\Phi(x)$. Then there is an additive polynomial $Q(x)$ in $K[x]$ with $\Phi(x) = Q(P(x))$.*

Proof. Suppose that $P = \sum_{i=I'}^{I} p_i \tau^i$ and $\Phi = \sum_{k=R'}^{R} \phi_k \tau^k$ satisfy $P(x) \mid \Phi(x)$. Then $R' \geq I'$. Our aim is to find $Q = \sum_{j=J'}^{J} q_j \tau^j$ with $\Phi = QP$ and q_j in K. Since $R' \geq I'$ we may assume that $I' = 0$. Suppose first that $R' = 0$, namely

Φ is separable. Since $P(x)$ divides $\Phi(x)$, $\ker P(x) = \{b \in \overline{K}; P(b) = 0\}$ is a subgroup of $\ker \Phi(x)$, and the quotient $\ker \Phi(x)/\ker P(x)$ is isomorphic to $P(\ker \Phi(x))$. The group $P(\ker \Phi(x))$ is an additive $\mathrm{Gal}(K_s/K)$-invariant subgroup of K_s. Hence there is some $c \neq 0$ in K such that $Q(x) = c \prod_u (x - u)$ (u in $P(\ker \Phi(x))$) is an additive polynomial (by part (ii) in the proof of Prop. 2.3) with $\Phi(x) = Q(P(x))$ and coefficients in K. In general, define ϕ'_k in $K^{p^{-R'}}$ by $\tau^{R'} \phi'_k = \phi_{k+R'} \tau^{R'}$ and $\Phi' = \sum_{k=0}^{R''} \phi'_k \tau^k$, where $R'' = R - R'$. Then $\Phi = \tau^{R'} \Phi'$, Φ' is separable and $P(x)$ divides $\Phi'(x)$. Hence there is $Q' = \sum q'_j \tau^j$ with $\Phi' = Q'P$. Then $Q = \tau^{R'} Q' = \sum q_j'^{p^{R'}} \tau^{j+R'}$ has coefficients in K, and it satisfies $\Phi = QP$, as required. $\square$

Recall that an elliptic functor E is the additive group equipped with an A-module structure.

Definition 3.5. (i) A morphism $P \neq 0$ in $\mathrm{Hom}(E, E')$ is called an *isogeny* (from E to E' or from φ to φ'). The functors E and E' (and φ, φ') are called *isogenous* if $\mathrm{Hom}(E, E') \neq \{0\}$. (ii) An isogeny $P = \sum_{i=0}^{m} b_i \tau^i$ in $K[\tau]$ is called *separable* if $b_0 \neq 0$. It is called *purely inseparable* if it is of the form $P = b\tau^m$ with some $b \neq 0$ in K and a positive integer m.

Remark 3.5. (i) If φ, φ' are isogenous then they have the same rank. (ii) If P is a purely inseparable isogeny then its kernel on $E(\overline{K})$ is $\{0\}$. (iii) If $m > 0$ then τ^m lies in $\mathrm{End}\,E$ if and only if φ is defined over $\mathbb{F}_{p^m}$. In this case the characteristic of φ is nonzero, and p^m is an integral power of $q_v = |A/v|$.

Any $P \neq 0$ in $\mathrm{End}_K \mathbb{G}_a$ can be written uniquely in the form $S\tau^j$, where S in $\mathrm{End}_K \mathbb{G}_a$ is separable and $j \geq 0$. For isogenies we have

Proposition 3.11. *Let P be an isogeny in $\mathrm{Hom}(E, E')$. If P is not separable then (i) char $\varphi = v \neq \{0\}$, and (ii) there are separable and purely inseparable isogenies S and $R = \tau^j$ such that $P = SR$ and $p^j = q_v^h$ for a positive integer h; here $q_v = |A/v| = |\mathbb{F}_v|$.*

Proof. Let $b\tau^j$ be the first term in P. The relation $P\varphi_a = \varphi'_a P$ implies that $i(a)^{p^j} = i(a)$ for all a in A. If char $\varphi = \{0\}$ then $j = 0$. If $j \neq 0$ then p^j has to be a power of $q_v = |A/v|$, where char $\varphi = v \neq 0$, as we now assume. Suppose that $P = \sum_{i=0}^{m} b_i \tau^{i+j}$, with $b_0 \neq 0$. Put $S = \sum_{i=0}^{m} b_i \tau^i$, and $R = \tau^j$. Then τ^j is an isogeny from φ to φ'', defined by $\varphi''_a = \sum a_i^{p^j} \tau^i$ if $\varphi_a = \sum a_i \tau^i$. Since $a_0^{p^j} = i(a)^{p^j} = i(a)$, φ'' is indeed an elliptic module. Since $K[\tau]$ is a domain, the relation $\varphi'_a SR = SR\varphi_a = S\varphi''_a R$ implies that $\varphi'_a S = S\varphi''_a$ for all a in A. Hence R lies in $\mathrm{Hom}(E, E'')$ and S in $\mathrm{Hom}(E'', E')$, as required. $\square$

Proposition 3.12. *Let $P \neq 0$ be an isogeny from φ to φ'. Then there exists an isogeny Q from φ' to φ, and $a \neq 0$ in A, such that $QP = \varphi_a$.*

Proof.   (i) We first consider the case where P is separable. Put $H = H_P$. Since $H(\overline{K})$ is a finite A-submodule of $E(\overline{K})$ there exists $a \neq 0$ in A with $\varphi_a(h) = 0$ for all h in $H(\overline{K})$. Since P is separable, the polynomial $P(x)$ divides $\varphi_a(x)$, and the ideal $(\varphi_a(x))$ lies in $(P(x))$. The universal property of quotients asserts that there is Q in $\operatorname{End} \mathbb{G}_{a,K}$ with $\varphi_a = QP$. It remains to show that Q is an isogeny. Since $P\varphi_b = \varphi_b' P$ for all b in A, we have

$$\varphi_b QP = \varphi_b \varphi_a = \varphi_a \varphi_b = QP\varphi_b = Q\varphi_b' P.$$

Since $K[\tau]$ is a domain we conclude that $\varphi_b Q = Q\varphi_b'$, as required.

(ii) If $P = \tau^j$, choose $a \neq 0$ in $\operatorname{char}\varphi$ (since P is purely inseparable, $\operatorname{char}\varphi \neq \{0\}$). Then $\varphi_a = \sum a_i \tau^i$ with $a_0 = 0$, and $\varphi_{a^j} = (\sum a_i \tau^i)^j = Q\tau^j$ for some Q.

(iii) Suppose that $P = SR$, where S is separable and R is purely inseparable, R in $\operatorname{Hom}(E, E'')$ and S in $\operatorname{Hom}(E'', E')$. Then there is T in $\operatorname{Hom}(E', E'')$ with $TS = \varphi_{a''}''$, and U in $\operatorname{Hom}(E'', E)$ with $UR = \varphi_{a'}$, for some a', a'' in A. Put $a = a'a''$ and $Q = UT$. Then $QP = UTSR = U\varphi_{a''}'' R = UR\varphi_{a''} = \varphi_{a'}\varphi_{a''} = \varphi_a$, as required. $\square$

Corollary 3.13. *If* $\operatorname{char}\varphi = 0$ *then* $\operatorname{End} E$ *is commutative.*

Proof. The map $D : \operatorname{End} E \to F$, $D(\sum a_i \tau^i) = a_0$, is a ring homomorphism. Proposition 3.12 asserts that for each $P \neq 0$ in $\operatorname{End} E$ there is $a \neq 0$ in A and Q in $\operatorname{End} E$ with $QP = \varphi_a$. Since $D(\varphi_a) = i(a) = a$, D is injective, and the corollary follows. $\square$

Let $E = E_\varphi$ be an elliptic functor over K. Then $E(\overline{K})$ is an A-module. We shall now determine those finite subgroups $H = H_P$ of $\mathbb{G}_{a,K}$ which are kernels of isogenies.

Proposition 3.14. *The morphism* $P \neq 0$ *in* $\operatorname{End}_K \mathbb{G}_a$ *is an isogeny from* E *to an elliptic functor* E' *if and only if* (i) $H(\overline{K})$ *is an* A-*submodule of* $E(\overline{K})$, (ii) $P = S\tau^j$ *where* S *is a separable isogeny, and* (1) $j = 0$ *if* $\operatorname{char}\varphi = \{0\}$, (2) p^j *is a power of* q_v *if* $\operatorname{char}\varphi = v \neq \{0\}$.

Proof. If P lies in $\operatorname{Hom}(E, E')$ then $P\varphi_a = \varphi_a' P$ for all a in A. If $P(b) = 0$ then $\varphi_a'(P(b)) = 0$. Hence $P(\varphi_a(b)) = 0$, whence (i). (ii) is verified in the proof of Prop. 3.11. In the opposite direction, we deal first with a separable P which satisfies (i) and (ii). By (i), the map $\psi_a = P\varphi_a$ in $\operatorname{End}_K \mathbb{G}_a$ is zero on $H(\overline{K})$, for every a in A. Since P is separable, $P(x)$ divides $\psi_a(x)$. Hence $(\psi_a(x))$ lies in $(P(x))$, and the universal property of quotients implies that there is φ_a' in $\operatorname{End}_K \mathbb{G}_a$ with $\psi_a = \varphi_a' P$, for any a in A. Then $\varphi' : A \to K[\tau]$ is the required elliptic module. In general $P = S\tau^j$, and it is verified in the proof of Prop. 3.11 that τ^j as in (ii(2)) lies in $\operatorname{Hom}(E, E'')$, where E'' is defined there. Since by (i) we have that $H_P(\overline{K})$ is an A-submodule of $E(\overline{K})$, it is clear that $H_S(\overline{K})$ is an A-submodule of $E''(\overline{K})$. Namely S is separable and satisfies (i). Hence it is an isogeny from E'' to some E' as shown above. The proposition follows. $\square$

Our next goal is to show that the torsion-free A-module $\operatorname{Hom}(E, E')$ is finitely generated and projective. Let m_∞ be the maximal ideal in the ring A_∞ of integers in the completion F_∞ of F at ∞.

Lemma 3.15. *For each nonnegative integer i, the group $A + m_\infty^i$ has finite index in F_∞.*

Proof. (i) If $B = \mathbb{F}_q[t]$ and $J = \mathbb{F}_q(t)$ then $J_\infty = \mathbb{F}_q((t^{-1}))$ and J_∞/B is compact. (ii) Take t in $A - \mathbb{F}_q$. The ring A is a finitely generated torsion-free module over the principal ideal domain $B = \mathbb{F}_q[t]$. Hence A is a free B-module of finite rank s. Choosing a basis we have $A = B^s$ and $F_\infty = A \otimes_B J_\infty = J_\infty^s$. By virtue of (i) the quotient J_∞/B is compact. Hence F_∞/A is compact. Since m_∞^i is open in F_∞, the quotient $F_\infty/(A + m_\infty^i)$ is finite, as required. $\qquad\square$

Proposition 3.16. *Let V be a subspace of $F \otimes_A \operatorname{Hom}(E, E')$ of finite dimension. Put $V_\infty = V \otimes_F F_\infty$ and $X = V \cap \operatorname{Hom}(E, E')$. Then (i) X is a discrete subset of V_∞; (ii) X is a finitely generated projective A-module.*

Proof. (i) The degree homomorphism $\deg : \operatorname{Hom}(E, E') \to \mathbb{Z}_{\geq 0}$ satisfies (1) $\deg u \geq 0$; (2) $\deg u = 0$ if and only if $u = 0$; (3) $\deg(au) = |a|^r \deg u$; (4) $\deg(u + v) \leq \max(\deg u, \deg v)$; and (5) $\deg(uv) = \deg u \cdot \deg v$ ($u, v \in V$; $a \in A$). By (3), $\deg$ extends to V_∞ on taking a in F_∞. Hence $\deg$ is a norm on V_∞, and X is discrete since X is a group and $\deg a \geq 1$ for all $a \neq 0$ in X. To prove (ii), let $x_1, \ldots, x_d$ be a basis of V over F which lies in X. We use this basis to identify V_∞ with $F_\infty \times \cdots \times F_\infty$ (d times), which we now denote by $[F_\infty]^d$. Hence we have $[A]^d \subset X \subset [F]^d \subset [F_\infty]^d$, and $[m_\infty^i]^d \subset [F_\infty]^d$, for all $i \geq 0$. Since X is discrete, there exists some $i \geq 0$ with $X \cap [m_\infty^i]^d = 0$. Hence X embeds in $[F_\infty/m_\infty^i]^d$, and $X/[A]^d$ in $[F_\infty/m_\infty^i + A]^d$, the latter being finite by Lemma 3.15. Since $[A]^d$ is a finitely generated A-module and $X/[A]^d$ is finite, X is a finitely generated A-module. Since X is torsion free over a Dedekind domain A, X is flat. Since X is also finitely generated we conclude that X is projective, as required. $\qquad\square$

Put $v = \operatorname{char} E$, and let w in $\operatorname{Spec} A$ be with $w \neq v$. For any V as above we have

Proposition 3.17. *The natural map from $X \otimes_A A_w$ to $\operatorname{Hom}_{A_w}(T_w E, T_w E')$ is injective.*

Proof. Since X is projective and finitely generated, we have $X \otimes_A A_w = \varinjlim_m X/(b^m)$ for any b in $A - \mathbb{F}_q$ whose only zero is at w. Since $T_w E$ and $T_w E'$ are free A_w-modules of rank r we have

$$\operatorname{Hom}_{A_w}(T_w E, T_w E') = \varprojlim_m \operatorname{Hom}_{A_w}(T_w E, T_w E')/(b^m).$$

Hence it suffices to show that the map $X/(b^m) \to \operatorname{Hom}_{A_w}(T_w E, T_w E')/(b^m)$ is injective for all m; we may take $m = 1$ on replacing b^m by b. Note that

$$\operatorname{Hom}_{A_w}(T_w E, T_w E')/(b) = \operatorname{Hom}_{A/(b)}(\ker \varphi_b, \ker \varphi_b').$$

Hence P in $\operatorname{Hom}(E, E')$ maps to zero in $\operatorname{Hom}_{A_w}(T_w E, T_w E')/(b)$ if and only if $P(a) = 0$ for any a in $\overline{K}$ with $\varphi_b(a) = 0$. Since b does not lie in v, we have $i(b) \neq 0$, and φ_b is separable. Hence φ_b divides P, and the universal property of quotients implies that there exists Q in $\mathrm{End}_K \mathbb{G}_a$ with $P = Q\varphi_b$. Since P is an isogeny from E to E', so is Q, as $K[\tau]$ is a domain. Hence P is zero in $X/(b)$, as required. $\quad\square$

Theorem 3.18. *Let r be the rank of E. Then (i) $\mathrm{Hom}(E, E')$ is a finitely generated projective A-module of rank bounded by r^2; (ii) for each $w \neq v = \mathrm{char}\, E$ in $\mathrm{Spec}\, A$, the natural map from $\mathrm{Hom}(E, E') \otimes_A A_w$ to $\mathrm{Hom}_{A_w}(T_w E, T_w E')$ is injective.*

Proof. Let V be a finite dimensional subspace of $F \otimes_A \mathrm{Hom}(E, E')$. Since $X = V \cap \mathrm{Hom}(E, E')$ is projective and finitely generated, the rank of X over A is equal to the rank of $X \otimes_A A_w$ over A_w (for all w), and the latter is bounded by r^2 by the proposition, since $T_w E$ is a free A_w-module of rank r. Since V is arbitrary, the rank of $\mathrm{Hom}(E, E')$ is bounded by r^2, and (i) as well as (ii) follow from the proposition. $\quad\square$

Proposition 3.19. (i) *The ring $F_\infty \otimes_A \mathrm{End}\, E$ is a division algebra of dimension at most r^2 over F_∞. (ii) The ring $F \otimes_A \mathrm{End}\, E$ is a division algebra of dimension $\leq r^2$ over F.*

Proof.   (i) By Theorem 3.18, $B_\infty = F_\infty \otimes_A \mathrm{End}\, E$ is an algebra of dimension $\leq r^2$ over F_∞. It has no zero divisors by the properties (1) and (2) of the degree homomorphism $\deg : (\mathrm{End}\, E) \otimes_A F_\infty \to \mathbb{Q}$ of Prop. 3.16. Given $b \neq 0$ in B_∞, let $L_b : B_\infty \to B_\infty$ the F_∞-linear operator $x \mapsto bx$. This B_∞ is a finite dimensional vector space over F_∞, and L_b is injective, since the ring B_∞ has no zero divisors. Hence L_b is surjective. Then there is $x \in B_\infty$ with $bc = 1$.

    (ii) Follows on replacing F_∞ by F in the proof of (i).
$\quad\square$

Corollary 3.20. *If $\mathrm{char}\, E = \{0\}$ then $(\mathrm{End}\, E) \otimes_A F_\infty$ is a field of degree $\leq r$ over F_∞.*

Proof. When $\mathrm{char}\, E = \{0\}$, Corollary 3.13 asserts that $\mathrm{End}\, E$ is commutative. Hence $F \otimes_A \mathrm{End}\, E$ is a commutative division algebra, that is, a field extension of F of degree $\leq r^2$. Theorem 3.18 implies that $F_w \otimes_A \mathrm{End}\, E$ embeds in the algebra $\mathrm{End}_{A_w}(T_w E) \otimes_{A_w} F_w$ of $r \times r$ matrices. Hence $F_w \otimes_A \mathrm{End}\, E$ is a direct sum of field extensions of F_w, the sum of whose degrees is bounded by r. Hence $[F \otimes_A \mathrm{End}\, E : F] \leq r$. Since $F_\infty \otimes_A \mathrm{End}\, E$ is a division algebra, ∞ does not split in $F \otimes_A \mathrm{End}\, E$, so $F_\infty \otimes_A \mathrm{End}\, E$ is a field of degree $\leq r$ over F_∞, as required. $\quad\square$

4. ELLIPTIC MODULES: GEOMETRIC DEFINITION

The definition in Chap. 3 of elliptic modules as A-structures on the additive group $\mathbb{G}_{a,K}$ over a field K over A has a natural generalization in which the field K, that is, the scheme $\operatorname{Spec} K$, is replaced by an arbitrary scheme S over A and $\mathbb{G}_{a,K}$ is replaced by an invertible (locally free rank one) sheaf $\mathbb{G}$ over S (equivalently a line bundle over S). An elliptic module of rank r over S will then be defined as an A-structure on $\mathbb{G}$ which becomes an elliptic module of rank r over K for any field K over S (thus $\operatorname{Spec} K \to S$). For our purposes it suffices to consider only affine schemes S and elliptic modules defined by means of a trivial line bundle $\mathbb{G}$ alone.

Definition 4.1. Let $S = \operatorname{Spec} B$ be an affine scheme over A; thus we have a ring homomorphism $i : A \to B$. Let $\mathbb{G}_{a,S} = \operatorname{Spec} B[x]$ be the additive group over S. (i) An *elliptic module* of rank r over S is a ring homomorphism $\varphi : A \to \operatorname{End} \mathbb{G}_{a,S} = B[\tau]$, $a \mapsto \varphi_a$, such that for any field K with a morphism $s : \operatorname{Spec} K \to S$ (or $s : B \to K$), the homomorphism $\varphi_s = s \circ \varphi : A \to B[\tau] \to K[\tau]$ is an elliptic module of rank r over K. (ii) Let E_φ be the functor from the category of rings over B (or affine schemes over S) to the category of A-modules, which associates to any ring R over B the additive group of R, together with the A-structure $a \circ r = \varphi_a(r)$ defined by φ. A functor E is called *elliptic* if it is isomorphic to some E_φ.

Remark 4.1. A morphism $P : E \to E'$ of elliptic functors $E = E_\varphi$ and $E' = E_{\varphi'}$ is P in $\operatorname{End} \mathbb{G}_{a,S} = B[\tau]$ with $P\varphi_a = \varphi'_a P$ for all a in A.

Let φ be an elliptic module over S. Then $\varphi_a = \sum_{i=0}^{I(a)} b_i(a)\tau^i$ for each a in A. Here the $b_i(a)$ lie in B, and $b_0(a) = i(a)$. Denote by $B^\times$ the group of units in B. Define v_p on $F^\times$ by $v_p(u\pi^m) = -m \log_p q$ where u is a unit and m an integer.

Lemma 4.1. *For each a in A we have that (i) $b_{rv_p(a)}(u)$ is a unit; (ii) if $i > rv_p(a)$, then $b_i(a)$ is nilpotent in B.*

Proof. (i) If $b_{rv_p(a)}(a)$ is not a unit then it lies in some maximal ideal m in B, and the reduction of φ modulo m cannot be an elliptic module of rank r over the residue field B/m; (i) follows. (ii) For each prime ideal $\mathfrak{P}$ in B, the reduction of φ modulo $\mathfrak{P}$ is an elliptic module over the fraction field of the domain $B/\mathfrak{P}$. Consequently for each $i > rv_p(a)$ we have that $b_i(a)$ lies in each prime ideal of B. Since the nilradical of B, which is defined to be the set of all nilpotent elements in B, is equal to the intersection of all prime ideals of B (see [AM], Prop. 1.8, p 5), (ii) follows. $\square$

Definition 4.2. (i) The element $f = \sum_0^n b_i\tau^i$ of $B[\tau]$ is called *strictly of degree* p^n if its leading coefficient b_n is a unit. (ii) An elliptic module φ over S is called *standard* if φ_a is strictly of degree $|a|^r$ for all a in A. Namely $b_i(a) = 0$ for $i > rv_p(a)$.

Y.Z. Flicker, *Drinfeld Moduli Schemes and Automorphic Forms: The Theory of Elliptic Modules with Applications*, SpringerBriefs in Mathematics, DOI 10.1007/978-1-4614-5888-3_4, © Yuval Z. Flicker 2013

Proposition 4.2. (1) *Every elliptic module is isomorphic to a standard module.* (2) *Every automorphism of a standard module is linear.*

Proof. We begin with two lemmas and then deduce the proposition.

Lemma 4.3. *Let $f = \sum_0^n b_i \tau^i$ be an element of $B[\tau]$ such that for some $d > 0$ we have that b_d is invertible and b_i is nilpotent for $i > d$. Then there exists $g = 1 + \sum_1^k c_j \tau^j$ in $B[\tau]$ such that the c_j are nilpotent (hence g is invertible) and $g f g^{-1}$ is strictly of degree d.*

Proof. We may assume that $n > d$. The ideal I generated by the nilpotent elements b_i $(d < i \le n)$ is nilpotent. Hence there is some $j > 1$ with $I^j = \{0\}$. In $B' = B/I^2$ we have $I^2 = 0$, and the degree of

$$f_1 = (1 - b_n b_d^{-p^{n-d}} \tau^{n-d}) f (1 - b_n b_d^{-p^{n-d}} \tau^{n-d})^{-1}$$

is less than n. Repeating this process with f_1 replacing f, where n is replaced by the exponent $n_1 < n$ of the highest power of τ in f_1, we obtain g_1 as in the lemma, such that $f' = g_1 f g_1^{-1}$ is strictly of degree d over B/I^2. Namely the coefficients b_i' $(d < i \le n)$ of f' lie in I^2. Since $I^j = \{0\}$ the lemma follows on repeating this process (with I replaced by I^2, I^4, etc.). $\qquad\square$

Lemma 4.4. *Suppose that $f_i = \sum_{j=0}^{d(i)} b_{ij} \tau^j$ $(i = 1, 2)$ and $h = \sum_{j=0}^d h_j \tau^j$ in $B[\tau]$ satisfy $f_2 h = h f_1$, h_d is either a unit or nilpotent $\ne 0$, $d(1) > 0$, and $b_{i,d(i)}$ are units. Then $d(1) = d(2)$ and h_d is a unit.*

Remark 4.2. The proof below shows that the same conclusion is valid if the assumption that "h_d is either a unit or nilpotent $\ne 0$" is replaced by "$h_d \ne 0$ and $\operatorname{Spec} B$ is connected," that is, B is not the direct product of two rings.

Proof. If h_d is nilpotent, then it lies in a maximal ideal m of B, and we may assume that h_d lies in m^{i-1} but not in m^i for some $i > 1$. We then replace B by B/m^i to have $m^i = 0$ and $h_d^2 = 0$. The highest term in $h f_1$ is $h_d b_{1,d(1)}^{p^d} \tau^{d+d(1)}$. Hence $d(2) \ge d(1)$, and $h_d b_{1,d(1)}^{p^d} = b_{2,d(1)} h_d^{p^{d(1)}}$. As $h_d^2 = 0$ and $b_{1,d(1)}$ is invertible, we deduce that $h_d = 0$. This contradiction implies that h_d is indeed a unit, and $d(1) = d(2)$. $\qquad\square$

Proof of proposition. Consider a in $A - \mathbb{F}_q$ and use (i) to produce g as in (i) such that $g \varphi_a g^{-1}$ is strictly of degree $|a|^r$. Then we define $\varphi_b' = g \varphi_b g^{-1}$ for all b in A. Since both φ and φ' are elliptic modules, the leading coefficients of φ_b and φ_b' are either units or nilpotents. Since $\varphi_a' \varphi_b' = \varphi_b' \varphi_a'$ for all b, it follows from (ii) that φ_b' is strictly of degree $|b|^r$. Hence φ' is a standard elliptic module. If P is an endomorphism of φ' and $\operatorname{Spec} B$ is connected, then it follows from (ii) that there is a nonnegative integer s such that P is strictly of degree s. If P is invertible then $s = 0$ and P is linear. The proposition follows. $\qquad\square$

Let $I \ne 0$ be an ideal in A. Let $E = E_\varphi$ be an elliptic functor of rank r over S.

Definition 4.3. Let $E_I = \operatorname{Ann} I | E$ be the subfunctor of E annihilated by I.

Corollary 4.5. *The annihilator E_I of I in E is finite and flat as an affine group scheme over S.*

Proof. Up to isomorphism we may assume that $E = E_\varphi$ is a standard elliptic functor. The affine group scheme underlying E is $\operatorname{Spec} B[x]$, and the subgroup underlying E_I corresponds to the ideal $J = J_I = \left(\sum_{j=0}^{rv_p(a)} b_j(a) x^{p^j} (a \in I) \right)$ in $B[x]$, namely it is $\operatorname{Spec} B[x]/J$. Since E is standard, the leading coefficients $b_{rv_p(a)}(a)$ are invertible in B. Let I' be an ideal in A with $I + I' = A$ and $I \cap I' = (a)$ for some $a \neq 0$ in A (cf. (ii) in the proof of Theorem 3.4). Then $B[x]/J_I \oplus B[x]/J_{I'} = B[x]/(\varphi_a(x))$ is finitely generated and free over B. Consequently $B[x]/J_I$ is finitely generated and projective, hence locally free and flat, of rank $|A/I|^r$ over B. $\qquad\square$

Let K be a field over B. Then there is a homomorphism $j : B \to K$ whose kernel m_K lies in $\operatorname{Spec} B$. Let m_B be an element of $\operatorname{Spec} B$. A finite flat scheme $\operatorname{Spec} C$ over $\operatorname{Spec} B$ is called *étale* over m_B if for any separably (equivalently, algebraically) closed field K over B with $m_K = m_B$ we have that $C \otimes_B K$ is a direct product of copies of K; the number of copies is equal to the rank of C over B.

Definition 4.4. Let $V(I)$ be the set of m in $\operatorname{Spec} A$ which contain the ideal I of A.

Proposition 4.6. *The affine group scheme underlying E_I is étale over $i^{-1}(\operatorname{Spec} A - V(I))$.*

Proof. Put $C = B[x]/J_I$. Fix any m_B in $\operatorname{Spec} B$ with $i(m_B) \not\supset I$. Since $i(m_B)$ is prime in A, $i(m_B) \not\supset I^d$ for all $d \geq 1$. The ideal class group of A is finite. Hence there is some $h \geq 1$ such that $I^h = (a)$ is principal, so that $i(a)$ does not lie in m_B. Let K be a separably closed field over B; thus we have $j : B \to K$, with $m_K = m_B$. For any $P(x)$ in $B[x]$ we apply $j : B \to K$ to the coefficients of $P(x)$ to obtain a polynomial $\overline{P}(x)$ in $K[x]$. Since $i(a)$ does not lie in $m_B = m_K$, $\overline{i(a)} = j(i(a)) \neq 0$, and $\overline{\varphi}_a(x)$ is separable. Hence the principal ideal $\overline{J}_I = (\overline{\varphi}_b; b \in I)$ in $K[x]$ is generated by a separable polynomial. But $\overline{\varphi}_a \in \overline{J}_I$, so $\overline{J}_I$ is generated by a separable polynomial. Consequently $C \otimes_B K = K[x]/\overline{J}_I$ is a direct product of copies of K, and C is étale over m_B for any m_B in $i^{-1}(\operatorname{Spec} A - V(I))$, as required. $\qquad\square$

Lemma 4.7. *Let $H \subset \operatorname{Spec} B[X]$ be a finite flat closed subscheme of rank r over $\operatorname{Spec} B$. Then the ideal I defining H is principal, generated by a uniquely determined monic polynomial h in $B[X]$ of degree r.*

Proof. Our assertion is equivalent to the following. Suppose $H = \operatorname{Spec} S$, $S = B[X]/I$, is finite and flat of rank r over B. Then $1, X, \ldots, X^{r-1}$ is a basis of S over B.

Consider the morphism $f : B^r \to S$, $f(a_0, \ldots, a_{r-1}) = \sum_{\{i; 0 \leq i < r\}} a_i X^i$. It is an isomorphism. To see this, we may pass to localizations. Thus we may assume that B is a local ring. Then S is a free B-module of rank r. So it suffices to show that f is surjective. By Nakayama's lemma, we reduce to the case of f modulo the maximal ideal of the local ring B, thus reducing to the case when B is field. In this case, it is clear that f is an isomorphism.

Now X^r is a linear combination $\sum_{\{i;0\leq i<r\}} a_i X^i$ of X^i, $0 \leq i < r$, with coefficients in B, modulo I. Thus $f = X^r - \sum_{\{i;0\leq i<r\}} a_i X^i$ is a monic element of I. Let $g \in B[X]$ be a nonzero element of I of lowest degree. The degree of g is at least r. If not, $1, X, \ldots, X^{r-1}$ are linearly dependent in S, that is, modulo I. Then there is a monomial $q \in B[X]$ such that $g - qf$ has lower degree than that of g, and $g - qf \in I$, since $g, f \in I$. Hence $g = qf$, and I is principal, generated by f. $\qquad\square$

Let $\varphi : A \to B[\tau]$ be an elliptic module of rank r over a ring B (or affine scheme $S = \operatorname{Spec} B$). Then $E_I(S)$, or $E_I(B)$, is the A/I-module

$$\operatorname{Hom}(S, E_I) = \operatorname{Hom}(B[x]/J_I, B) = \{b \in B; \varphi_a(b) = 0 \text{ for all } a \in I\}.$$

Definition 4.5.　　(1) A *structure of level* I on the elliptic functor $E = E_\varphi$ of rank r over $S = \operatorname{Spec} B$ is an A-module homomorphism $\psi :$ $(I^{-1}/A)^r \to E_I(S) = \operatorname{Spec} B[x]/J_I$ such that $J_I = (P(x))$, $P(x) = \prod_{u \in (I^{-1}/A)^r} (x - \psi(u))$. Then E_I is the kernel of the polynomial map $P(x) = \prod_{u \in (A/I)^r} (x - \psi(u))$, $P : \mathbb{G}_a \to \mathbb{G}_a$. In other words, $\sum_{u \in (A/I)^r} (\psi(u))$ is E_I as divisors on E.

(2) Let $\mathbb{F}_{r,I}$ be the covariant functor from the category $\mathbf{R}_A$ of rings B over A to the category $\mathbf{Set}$ of sets, which assigns to the ring B the set of isomorphism classes of elliptic modules $\varphi : A \to B[\tau]$ over B of rank r with a structure $\psi : (I^{-1}/A)^r \to E_I(B)$ of level I. The functor $\mathbb{F}_{r,I}$ is anti-equivalent to a contravariant functor, also denoted by $\mathbb{F}_{r,I}$, from the category $\mathbf{S}_A$ of affine schemes over A to $\mathbf{Set}$, defined by $\mathbb{F}_{r,I}(S) = \mathbb{F}_{r,I}(B)$ if $S = \operatorname{Spec} B$.

(3) A covariant functor $\mathbb{F} : \mathbf{C} \to \mathbf{Set}$ is called *representable* if there is an object C in the category $\mathbf{C}$ such that $\mathbb{F}(B) = \operatorname{Hom}(C, B)$ for all B in $\mathbf{C}$. A contravariant functor $\mathbb{F} : \mathbf{C} \to \mathbf{Set}$ is called representable if there is T in $\mathbf{C}$ with $\mathbb{F}(S) = \operatorname{Hom}(S, T)$ for all S in $\mathbf{C}$.

Remark 4.3. It is illuminating to describe the (scheme $\mathbb{G}_{a,B} = \operatorname{Spec} B[x]$ over $S = \operatorname{Spec} B$ which underlies the) functor E by a diagram (Fig. 1).

The sections $\psi(u)$ may intersect only over the support $i^{-1}(V(I))$ of I in $S = \operatorname{Spec} B$. If the image of S in $\operatorname{Spec} A$ does not intersect $V(I)$ then E_I is étale over S, namely the sections $\psi(u) : S \to E_I$ do not intersect, and the graph is as in Fig. 2.

Our aim in Theorem 4.10 is to show that the functor $\mathbb{F}_{r,I}$ is representable, for any ideal $I \neq 0$ in A with $[V(I)] \geq 2$.

Let φ be an elliptic module of rank r over B with structure $\psi : (I^{-1}/A)^r \to E_I(S) \subset B$ of level $I \neq 0$. Suppose that m in $\operatorname{Spec} A$ contains I. Let m_B be a maximal ideal in B. Fix $u \neq 0$ in $(m^{-1}/A)^r$. Recall that we have $i : A \to B$.

Lemma 4.8. *If $\psi(u)$ lies in m_B then m_B contains the image $i(m)$ of m.*

Proof. Denote the field B/m_B by $\overline{B}$. Let $\overline{\varphi} : A \to \overline{B}[\tau]$ and $\overline{\psi} : (I^{-1}/A)^r \to \overline{B}$ be the elliptic module with level structure over $\overline{B}$ obtained on composing φ, ψ

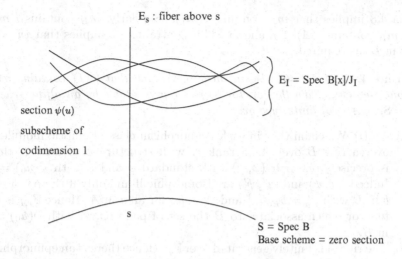

E_s : fiber above s

$E_I = \text{Spec } B[x]/J_I$

section $\psi(u)$

subscheme of

codimension 1

s

S = Spec B
Base scheme = zero section

FIGURE 1. The scheme $\mathbb{G}_{a,B}$

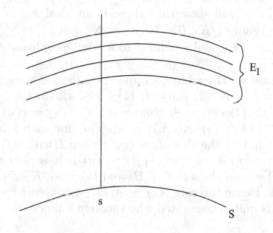

E_I

s

S

FIGURE 2. The étale case

with $B \to \overline{B}$. Suppose that $\psi(u)$ lies in m_B, namely $\overline{\psi}(u) = 0$ in $\overline{B}$. If $i(m)$ does not lie in m_B then there is a in m with $\overline{i(a)} \neq 0$. Hence $\overline{\varphi}_a$ is separable, and $\overline{J}_m = (\overline{\varphi}_b(x);\ b$ in $m)$ is generated by a separable polynomial in $\overline{B}[x]$. Hence the restriction $\overline{\psi} : (m^{-1}/A)^r \to \overline{B}$ of the level structure $\overline{\psi}$ to $(m^{-1}/A)^r$ is injective. This contradicts the assumption that $\overline{\psi}(u) = 0$ for $u \neq 0$. Hence $i(m)$ lies in m_B. $\qquad \square$

Suppose that there are $m_1 \neq m_2$ in $V(I)$. Fix $u_j \neq 0$ in $(m_j^{-1}/A)^r$.

Proposition 4.9. *Put $u_0 = u_1 + u_2$. Then $\psi(u_0)$ is a unit in B.*

Proof. Let a be an element of A in m_2 such that $a - 1$ lies in m_1. If $\psi(u_0)$ is not a unit in B, then it lies in some maximal ideal m_B of B. Since $au_0 = u_1 + (1 - a)u_1 + au_2$, we have that $\psi(u_1) = \psi(au_0) = \varphi_a\psi(u_0)$ lies in m_B.

Lemma 4.8 implies that m_B contains $i(m_1)$. Similarly, m_B contains $i(m_2)$; hence m_B contains $i(A)$, 1_B, and B. This contradiction implies that $\psi(u_0)$ is a unit in B, as required. $\qquad \square$

Theorem 4.10. *Let $I \neq 0$ be an ideal in A such that $V(I)$ contains more than one element. Then the functor $\mathbb{F}_{r,I}$ is representable by an affine scheme $M_{r,I} = \operatorname{Spec} A_{r,I}$ of finite type over A.*

Proof. (i) We claim that in each isomorphism class of elliptic modules φ over a ring B over A, of rank r, with structure ψ of level I, there is precisely one pair (φ, ψ) with standard φ and ψ with $\psi(u_0) = 1$. Indeed, (φ, ψ) and (φ', ψ') are isomorphic if and only if there is a unit b in B with $\varphi'_a = b\varphi_a b^{-1}$ and $\psi' = b\psi$ for all a in A. Hence $\mathbb{F}_{r,I}$ is the functor which associates to B the set of pairs (φ, ψ) with $\psi(u_0) = 1$ in B.

(ii) The ring A is finitely generated over $\mathbb{F}_q$. Hence there is an epimorphism $e : T = \mathbb{F}_q[t_1, \ldots, t_s] \to A$, with kernel K. Let A_1 be the free ring $A[b_{ij}, b_i, c(u)]$; here $1 \leq i \leq s$, $0 \leq j \leq rv_q(e(t_i))$, u ranges over $(I^{-1}/A)^r$. We will show that there is an ideal K_1 in A_1 such that $\mathbb{F}_{r,I}(B) = \operatorname{Hom}(A_1/K_1, B)$. To define K_1, put $\varphi_{t_i} = \sum_j b_j(t_i)\tau^{j \log_p q}$, where $b_j(t_i) = b_{ij}$, and extend φ to a homomorphism $\varphi : T \to A_1[\tau]$, $t \mapsto \varphi_t = \sum_i b_i(t)\tau^{i \log_p q}$ $(0 \leq i \leq rv_q(e(t)))$. Let K_1 be the ideal in A_1 generated by the relations implied by the following identities: (1) $\varphi_t = 0$ for t in K; (2) $b_{i0} = t_i$ $(1 \leq i \leq s)$; (3) $b_i b_{rv_q(e(t_i))}(t_i) = 1$ $(1 \leq i \leq s)$; (4) $c(0) = 0$, $c(u_0) = 1$, $c(u + v) = c(u) + c(v)$ for all u, v in $(I^{-1}/A)^r$; (5) $\varphi_t(c(u)) = c(e(t)u)$ for each t in T and u in $(I^{-1}/A)^r$; and (6) the ideal $J_I = (\varphi_t(x); t$ in T with $e(t)$ in $I)$ in $A_1[x]$ (here $\tau(x) = x^p$) is equal to $(\prod_u (x - c(u)))$. It is clear that there is a bijection between the sets $\mathbb{F}_{r,I}(B)$ and $\operatorname{Hom}(A_1/K_1, B)$ for every ring B over A. Hence the ring $A_{r,I} = A_1/K_1$ represents the functor $\mathbb{F}_{r,I}$. Since A_1 is finitely generated, the theorem follows. $\qquad \square$

Definition 4.6. The elliptic module $\varphi : A \to A_{r,I}[\tau]$ with level structure $\psi : (I^{-1}/A)^r \to E_I(A_{r,I})$ in $\mathbb{F}_{r,I}(A_{r,I}) = \operatorname{Hom}(A_{r,I}, A_{r,I})$, which corresponds to the identity map $A_{r,I} \to A_{r,I}$, is called the *universal* elliptic module with a structure of level I. It has the form $\varphi_a = \sum_i \bar{b}_i(a)\tau^{i \log_p q}$ and $\psi(u) = \bar{c}(u)$ for a in A and u in $(I^{-1}/A)^r$, where $\bar{b}_i(a)$ and $\bar{c}(u)$ denote the image of $b_i(a)$ and $c(u)$ in $A_{r,I} = A_1/K_1$.

Lemma 4.11. *Let (φ, ψ) be the universal elliptic module of rank r with structure of level I. Then the map $\psi : (I^{-1}/A)^r \to A_{r,I}$ is injective.*

Proof. We have to show that if $u \neq 0$ in $(I^{-1}/A)^r$ then $\bar{c}(u) \neq 0$ in $A_{r,I}$. If $\bar{c}(u) = 0$ then relation (6) implies that x^2 divides $\varphi_a(x)$ for all a in I. But this is impossible since the homomorphism $b_0 : A \to A_{r,I}$ is injective, and $\varphi_a(x) = b_0(a)x +$ terms of higher degree. $\qquad \square$

Lemma 4.12. *Let $\varphi : A \to B[\tau]$ be an elliptic module, $I' \supset I \neq 0$ ideals in A with $[V(I)] \geq 2$, and $\psi : (I^{-1}/A)^r \to E_I(B)$ an I-level structure. Then the restriction of ψ to $(I'^{-1}/A)^r$ is an I'-level structure on the elliptic module φ.*

Proof. It suffices to prove the lemma when $B = A_{r,I}$ and (φ, ψ) is the universal elliptic module with an I-level structure. The ideal $J_{I'}$ divides the ideal J_I, and (by (6)) J_I is generated by the separable polynomial $\prod_u (x - \psi(u))$, u in $(I^{-1}/A)^r$. The relation (5) in (ii) of the proof of Theorem 4.10 implies that for each u in $(I'^{-1}/A)^r$ we have that $\varphi_a(\psi(u)) = 0$, for every a in I'; moreover, for each u in $(I^{-1}/A)^r - (I'^{-1}/A)^r$ there is a a in $I' - I$ such that $\varphi_a(\psi(u)) \neq 0$. Hence $J_{I'} = (\prod_u (x - \psi(u)); u \in (I'^{-1}/A)^r)$, and the lemma follows. $\qquad\square$

Definition 4.7. A homomorphism $B \to B'$ of rings is called *separable* if every element of B' is separable over the image of B.

Proposition 4.13. *Let $I' \supset I \neq 0$ be two ideals in A with $[V(I')] \geq 2$. Then the natural map $A_{r,I'} \to A_{r,I}$ is separable, finite, and flat.*

Proof. The ring $A_{r,I}$ is generated over $A_{r,I'}$ by the elements $c(u)$, where u ranges over $(I^{-1}/A)^r - (I'^{-1}/A)^r$. These $c(u)$ satisfy the relations (4), (5), and (6) in (ii) of the proof of Theorem 4.10. In particular, for any $a \neq 0$ in I, (5) implies that

$$0 = c(au) = \varphi_a(c(u)) = \sum_{i=0}^{rv_q(a)} b_i(a) c(u)^{q^i}.$$

Since $b_0 : A \to A_{r,I}$ is injective, the map $A_{r,I'} \to A_{r,I}$ is separable. The relation (3) implies that the leading coefficient $b_{rv_q(a)}(a)$ is invertible in A_1. As in the proof of Corollary 4.5, this implies that $A_{r,I}$ is a flat $A_{r,I'}$-module of finite rank. The rank is bounded by the product of the number of generators $c(u)$ (which is bounded by $|A/I|^r - |A/I'|^r$), and $|a|^r$, for any $a \neq 0$ in I. The proposition follows. $\qquad\square$

Definition 4.8. (i) Put $A_r = \varinjlim A_{r,I}$; the direct limit is defined by means of the transition maps $A_{r,I'} \to A_{r,I}$ for $I' \supset I \neq 0$. (ii) Put $M_r = \operatorname{Spec} A_r = \varprojlim M_{r,I}$, where $M_{r,I} = \operatorname{Spec} A_{r,I}$. (iii) Let $\mathbb{F}_r$ be the functor $\varprojlim \mathbb{F}_{r,I}$, which associates to the ring B over A the set $\varprojlim \mathbb{F}_{r,I}(B) = \varprojlim \operatorname{Hom}(A_{r,I}, B)$. The transition maps $\mathbb{F}_{r,I}(B) \to \mathbb{F}_{r,I'}(B)$ are given by mapping (φ, ψ) to (φ, ψ'), where ψ' is the restriction of $\psi : (I^{-1}/A)^r \to E_\varphi(B)$ to $(I'^{-1}/A)^r$.

Corollary 4.14. *(i) The functor $\mathbb{F}_r$ is represented by the ring A_r (and scheme M_r). (ii) The functor $\mathbb{F}_r$ associates to the scheme $S = \operatorname{Spec} B$ over A the set of isomorphism classes of elliptic modules φ over B of rank r with a structure of all levels, namely an A-module homomorphism $\psi : (F/A)^r \to E(S) = B$ such that for any ideal $I \neq 0$ in A the restriction $\psi : (I^{-1}/A)^r \to E_I(S)$ of ψ to $(I^{-1}/A)^r$ is a structure of level I.*

Proof. This follows from the definitions. $\qquad\square$

Let $\mathbb{A}$ denote the ring of adèles of F. It consists of all sequences $(x_w; w \in |C|)$ with x_w in F_w for all closed points w of C and x_w in A_w for almost all w.

Here $|C|$ is the set of closed points of C. Addition and multiplication are componentwise. The topology on $\mathbb{A}$ is defined by taking a fundamental system of (open) neighborhoods of the identity to consist of $\prod_{w \in V} N_w \times \prod_{w \notin V} A_w$, where V ranges over all finite subsets of the set $|C|$, and N_w is a (open) neighborhood of zero in F_w. Then $\mathbb{A}$ is a locally compact ring.

For any $r \geq 1$, let $\mathrm{GL}(r, F_w)$ denote the topological group of $r \times r$ invertible matrices with entries in F_w and $\mathrm{GL}(r, A_w)$ the group of $r \times r$ matrices with entries in A_w and determinant in $A_w^{\times}$. By $\mathrm{GL}(r, \mathbb{A})$ denote the group of sequences $(g_w$ in $\mathrm{GL}(r, F_w); w$ in $|C|)$, with g_w in $\mathrm{GL}(r, A_w)$ for all but finitely many points w.

The topology on $\mathrm{GL}(r, \mathbb{A})$ is defined by a system $\prod_{w \in V} N_w \times \prod_{w \notin V} \mathrm{GL}(r, A_w)$ of neighborhoods of the identity, where V is finite and N_w is a neighborhood of 1 in $\mathrm{GL}(r, F_w)$. Then $\mathrm{GL}(r, \mathbb{A})$ is a locally compact group. The group $\mathrm{GL}(1, \mathbb{A})$ is called the group of idèles; it is denoted by $\mathbb{A}^{\times}$, being the multiplicative group of $\mathbb{A}$. The ring of finite adèles $\mathbb{A}_f$ and the groups $\mathrm{GL}(r, \mathbb{A}_f)$ are defined analogously on replacing the set C by the set $\mathrm{Spec}\, A = C - \{\infty\}$, namely on omitting the component at the place ∞. The set F is dense in $\mathbb{A}_f$, $F^{\times}$ is discrete in $\mathbb{A}_f^{\times}$, and $F^{\times} \backslash \mathbb{A}_f^{\times}$ is compact.

Let $\widehat{A}$ denote the compact open subring $\prod_w A_w$ (w in $\mathrm{Spec}\, A$) of $\mathbb{A}_f$. Since $(F/A)^r = \prod_w (F_w/A_w)^r$, the group $\mathrm{GL}(r, \widehat{A}) = \prod_w \mathrm{GL}(r, A_w)$ is equal to the group $\mathrm{Aut}_A(F/A)^r$ of automorphisms of $(F/A)^r$. The matrix g in $\mathrm{GL}(r, \widehat{A})$ is the automorphism of multiplication by g. Hence $\mathrm{GL}(r, \widehat{A})$ acts on $(F/A)^r$ and on the functor $\mathbb{F}_r$, by mapping (φ, ψ) to $(\varphi, \psi \circ g)$.

Proposition 4.15. *There is an action of* $\mathrm{GL}(r, \mathbb{A}_f)$ *on* $\mathbb{F}_r$ *extending the action of* $\mathrm{GL}(r, \widehat{A})$.

Proof.   (i) Let $M(r)$ denote the algebra of $r \times r$ matrices. The semigroup $G_1 = \mathrm{GL}(r, \mathbb{A}_f) \cap M(r, \widehat{A})$ has the property that $\mathrm{GL}(r, \mathbb{A}_f) = F^{\times} G_1$. We shall define an action of G_1 on $\mathbb{F}_r$ which is trivial on $F^{\times} \cap G_1 = A - \{0\}$, thereby defining an action of $\mathrm{GL}(r, \mathbb{A}_f)$ on $\mathbb{F}_r$ which is trivial on $F^{\times}$. Thus, for each g in G_1 and a ring B over A, we have to construct a map $\rho(g) : \mathbb{F}_r(B) \to \mathbb{F}_r(B)$, such that $\rho(gg') = \rho(g)\rho(g')$ for all g, g' in G_1. Let $\varphi : A \to B[\tau]$ be an elliptic module of rank r over B, with a level structure $\psi : (F/A)^r \to E(S) = B$. We shall now construct a new pair $(\rho(g))(\varphi, \psi) = (\varphi' = g\varphi, \psi' = g\psi)$. It suffices to deal—as we now do—with the case where $B = A_r$, and (φ, ψ) is the universal elliptic module with a structure of all levels.

(ii) Multiplication by g defines an epimorphism $g : (F/A)^r \to (F/A)^r$ with a finite kernel H. Consider the polynomial $P_H(x) = \prod_{h \in H}(x - \psi(h))$ in $B[x]$. Let H' denote the kernel of the restriction $\psi|H$ of ψ to H. Put $H'' = H/H'$. Then $\psi|H$ factorizes through $\psi'' : H'' \to B$, and $P_H(x) = \prod_{h \in H''}(x - \psi''(h))^{|H'|}$. The values $\psi''(h)$ are distinct. Since $aH \subset H$ for each a in A, we have that $P_H(\varphi_a(b)) = 0$ for any b in B with $P_H(b) = 0$. Hence $P_H(x)$ divides $P_H(\varphi_a(x)) = \prod_{h \in H''}(\varphi_a(x) - \psi''(h))^{|H'|}$.

Lemma 4.16. *Let in as here $P(x)$, $\Phi(x)$ be additive polynomials in $B[x]$ such that* (1) *$P(x)$ divides $\Phi(x)$;* (2) *$P(x)$ is separable; and* (3) *$\Phi(x)$ splits as a product of linear factors over B. Then there is an additive polynomial $Q(x)$ in $B[x]$ with $\Phi(x) = Q(P(x))$.*

Proof. This is the same as the proof of Lemma 3.10. □

Applying Lemma 4.16 with $P(x) = P_H(x)$ and $\Phi(x) = P_H(\varphi_a(x))$, we conclude that there exists an additive polynomial $\varphi_a'(x)$ in $B[x]$ with φ_a' $(P_H(x)) = P_H(\varphi_a(x))$. It is clear that $\varphi' : A \to B[\tau]$, $a \mapsto \varphi_a'$, is an elliptic module of rank r over B. Put $E' = E_{\varphi'}$, and $g\varphi = \varphi'$. Then there is a map $\psi' : (F/A)^r \to E'(S) = B$ which makes commutative the diagram

$$
\begin{array}{ccccccccc}
0 & \to & H & \to & (F/A)^r & \overset{g}{\to} & (F/A)^r & \to & 0 \\
 & & \psi\downarrow\wr & & \psi\downarrow & & \downarrow\psi' & & \\
0 & \to & \psi(H) & \to & E(B) & \underset{P_H}{\to} & E'(B) & \to & 0.
\end{array}
$$

Since ψ is injective, so is ψ'; hence ψ' is a level structure. Put $g\psi$ for ψ'. Since (φ, ψ) is the universal elliptic module, for each ring B over A, we can associate to any g in G_1 the map $\rho(g) : \mathbb{F}_r(B) \to \mathbb{F}_r(B)$ which takes the elliptic module φ with level structure ψ to the pair $(\varphi' = g\varphi, \psi' = g\psi)$. Since $\rho(gg') = \rho(g)\rho(g')$ for all g, g' in G_1, this defines an action of G_1 on the functor $\mathbb{F}_r$.

(iii) If $g = a$ is a scalar matrix in $A - \{0\}$, then the definition of the level structure ψ implies that $(P_{II}(x)) = (\varphi_a(x))$ in $B[x]$. Hence there is b in $B^\times$ with $\varphi_a(x) = bP_H(x)$. Since $\varphi_a\varphi_{a'} = \varphi_{a'}\varphi_a$ for any a' in A, we have $P_H\varphi_a = b^{-1}\varphi_a b P_H$ for every a in A. Hence $\varphi_a' = b^{-1}\varphi_a b$, and the elliptic module φ' is isomorphic to φ. Hence the elements of $A - \{0\}$ act trivially on the functor $\mathbb{F}_r$, and we obtain an action of $\mathrm{GL}(r, \mathbb{A}_f)/F^\times$ on $\mathbb{F}_r$; the proposition follows.

□

Remark 4.4. The action of $\mathrm{GL}(r, \mathbb{A}_f)/F^\times$ on $\mathbb{F}_r$ induces an action of this group on the ring A_r and the scheme $M_r = \mathrm{Spec}\, A_r$.

Let $I \neq 0$ be an ideal in A.

Definition 4.9. (*i*) The congruence subgroup U_I of $\mathrm{GL}(r, \widehat{A})$ is the group of all g in $\mathrm{GL}(r, \widehat{A})$ with $g - 1$ in $M(r, I\widehat{A})$. (*ii*) If U is a group which acts on a ring B then we write B^U for the ring of b in B fixed by U. (*iii*) If U is a group which acts on an affine scheme M then we write $U\backslash M$ for the quotient of M by the action of U.

Remark 4.5. (i) U_I is an open compact subgroup of $\mathrm{GL}(r, \widehat{A})$. (ii) $U_I\backslash M_r$ is an affine scheme which is equal to $\mathrm{Spec}\, A_r^{U_I}$.

Definition 4.10. A domain B is called *normal* if it is integrally closed in its fraction field, equivalently, if its localization $B_{\mathfrak{P}}$ at each of its prime ideals $\mathfrak{P}$ is an integrally closed domain, that is, normal. A ring B is *normal* if $B_{\mathfrak{P}}$ is a normal domain for every prime ideal $\mathfrak{P}$ of B. A Noetherian normal ring is a finite direct product of normal domains.

Lemma 4.17. *Let $B \to C$ be a finite separable homomorphism of rings over $\mathbb{F}_p$. Suppose the ring B is normal and $(*)\operatorname{Hom}(C, K) \to \operatorname{Hom}(B, K)$ is an isomorphism for every algebraically closed field K. Then $B \to C$ is an isomorphism.*

Proof. Without loss of generality we may assume that B and C are domains. Let B', C' denote the fraction fields of B, C. Since $B' \to C'$ is separable, $(*)$ implies that $B' \to C'$ is an isomorphism. In particular, $B \to C$ is injective. Since $B \to C$ is finite, C is integral over B. Hence C lies in the integral closure of B in B'. Since B is normal we have that $B = C$, as required. $\square$

Remark 4.6. In the next proposition we use the fact that for $I \neq 0$ with $[V(I)] \geq 2$, the ring $A_{r,I}$ is normal. The proof, which involves techniques independent of Section 4 and Chaps. II–IV, is given in [D1], Sections 1, 4, and 5C).

Proposition 4.18. *The ring $A_{r,I}$ is equal to $A_r^{U_I}$ for $I \neq 0$ with $[V(I)] \geq 2$. Equivalently, the scheme $U_I \backslash M_r = \operatorname{Spec} A_r^{U_I}$ is equal to $M_{r,I}$.*

Proof. We have to show that $A_{r,I} = A_{r,J}^{U_I}$ for any $0 \neq J \subset I$ in A. In view of Lemma 4.17 it suffices to show that for any algebraically closed field K the natural map $\operatorname{Hom}(A_{r,J}^{U_I}, K) \to \operatorname{Hom}(A_{r,I}, K)$ is an isomorphism. This map takes the elliptic module $\varphi : A \to K[\tau]$ with a structure ψ of level J, to the pair (φ, ψ'), where ψ' is the restriction of ψ to $(I^{-1}/A)^r$. To construct an inverse to this map we need to extend a structure ψ' of level I to a structure ψ of level J. This can be done since K is algebraically closed, and the proposition follows. $\square$

For any v in $\operatorname{Spec} A$ we make the following:

Definition 4.11. (1) Let $M_{r,v} = M_r \times_{\operatorname{Spec} A} \operatorname{Spec}(A/v)$ be the fiber of M_r at v. (2) Let $M_{r,v,I} = M_{r,I} \times_{\operatorname{Spec} A} \operatorname{Spec}(A/v) = (U_I \backslash M_r) \times_{\operatorname{Spec} A} \operatorname{Spec}(A/v) = U_I \backslash M_{r,v}$ be the fiber of $M_{r,I}$ at v.

Remark 4.7. The affine scheme $M_{r,v}$ represents the functor $\mathbb{F}_{r,v}$, which is the restriction of the functor $\mathbb{F}_r$ to the category of rings B over A/v, namely rings B over A with characteristic v. In particular, the set $\mathbb{F}_{r,v}(B)$ consists of pairs (φ, ψ), where φ is an elliptic module *of characteristic v*.

5. COVERING SCHEMES

The group $\mathrm{GL}(r, \mathbb{A}_f)$ acts (by Prop. 4.15) on the moduli scheme $M_r = \mathrm{Spec}\, A_r = \varprojlim M_{r,I}$ constructed in Theorem 4.10. The central group $F^\times$ acts trivially. In this section we construct a covering scheme $\widetilde{M}_r$ of M_r for which the action of $\mathrm{GL}(r, \mathbb{A}_f)$ extends nontrivially to an action of $(\mathrm{GL}(r, \mathbb{A}_f) \times D_\infty^\times)/F^\times$, where D_∞ is a division algebra of rank r over F_∞.

As in Chap. 2 we have F, ∞, A, and we let F_∞ be the completion of F at ∞, A_∞ the ring of integers in F_∞, m_∞ the maximal ideal of A_∞, $\pi = \pi_\infty$ a local uniformizer, $q = q_\infty$ the cardinality of A_∞/m_∞, $|.| = |.|_\infty$ the absolute value on F_∞ which is normalized by $|\pi| = q^{-1}$ (equivalently $|a| = |A/aA|$ for $a \neq 0$ in A). Put $v_p(a) = \log_p |a|$ for $a \in F_\infty^\times$. Then $v_p(\pi) = -\log_p q$. Note that $A \cap A_\infty$ consists of the constant functions on C (thus $A \cap A_\infty = \mathbb{F}_q$ if $F = \mathbb{F}_q(C)$ as in Chap. 2). Moreover, F_∞ is $\mathbb{F}_q((\pi))$. As usual, given a ring B of characteristic p we denote by $B[\tau]$ the ring generated by τ over B subject to the relations $\tau b = b^p \tau$ for all b in B.

A reduced ring $\mathbb{B}$ is called *perfect* if for each b in $\mathbb{B}$ there exists a (necessarily unique) x in $\mathbb{B}$ with $x^p = b$. Equivalently, the Frobenius homomorphism from $\mathbb{B}$ to $\mathbb{B}$, by $x \mapsto x^p$, is an isomorphism. Let $\mathbb{B}$ be a perfect ring. Then for each b in $\mathbb{B}$ there exists a unique $b^{p^{-1}}$ in $\mathbb{B}$.

Definition 5.1. (1) Let $\mathbb{B}((\tau^{-1}))$ be the ring of formal Laurent series $\sum_{i=N}^{-\infty} b_i \tau^i$ with coefficients b_i in $\mathbb{B}$, N in $\mathbb{Z}$, and multiplication

$$\left(\sum_i b_i' \tau^i\right)\left(\sum_j b_j \tau^j\right) = \sum_{i,j} b_i' b_j^{p^i} \tau^{i+j}.$$

(2) Let $\mathbb{B}[[\tau^{-1}]]$ be the subring of formal Taylor series in τ^{-1} (thus $N \leq 0$).

(3) The ring $\mathbb{B}((\tau^{-1}))$ is a topological ring, where a system of open neighborhoods of 0 is given by $\tau^{-j}\mathbb{B}[[\tau^{-1}]]$ ($j \geq 0$).

Let $\varphi : A \to \mathbb{B}[\tau]$ be a standard elliptic module of rank r over a perfect ring $\mathbb{B}$ over A.

Lemma 5.1. *The elliptic module φ extends uniquely to a continuous ring homomorphism $\varphi : F_\infty \to \mathbb{B}((\tau^{-1}))$.*

Proof. The homomorphism φ extends uniquely to a homomorphism $\varphi : F \to \mathbb{B}((\tau^{-1}))$ such that for each a in F we have $\varphi_a = \sum_{i=N}^{-\infty} b_i(a)\tau^i$, where $N = rv_p(a)$ and $b_N(a)$ is invertible. In particular, for each $t \geq 0$, the image of $F \cap m_\infty^t$ lies in $\tau^{-t}\mathbb{B}[[\tau^{-1}]]$ (when $t = 0$, we put $m_\infty^t = m_\infty^0$ for A_∞). Hence φ extends to F_∞ by continuity, and the lemma follows. $\square$

Corollary 5.2. *The ring $\mathbb{B}$ contains a copy of $\mathbb{F}_q$.*

Proof. We have $F_\infty = \mathbb{F}_q((\pi))$, and $\mathbb{F}_q = A_\infty/m_\infty$ embeds in $\mathbb{B}[[\tau']]/\tau'\mathbb{B}[[\tau']] = \mathbb{B}$, where $\tau' = \tau^{rv_p(\pi)}$. $\square$

Y.Z. Flicker, *Drinfeld Moduli Schemes and Automorphic Forms: The Theory of Elliptic Modules with Applications*, SpringerBriefs in Mathematics, DOI 10.1007/978-1-4614-5888-3_5, © Yuval Z. Flicker 2013

As usual, we denote the algebraic closure of $\mathbb{F}_p$ by $\overline{\mathbb{F}}_p$.

Definition 5.2. Let $\varphi' : F_\infty \to \overline{\mathbb{F}}_p((\tau^{-1}))$ be the ring homomorphism over $\mathbb{F}_q$ defined by $\varphi'_{\boldsymbol{\pi}} = \tau^{rv_p(\boldsymbol{\pi})}$.

Remark 5.1. (1) Recall that $v_p(\boldsymbol{\pi}) = -\log_p q$. (2) Since $F_\infty = \mathbb{F}_q((\boldsymbol{\pi}))$, φ' is uniquely determined by its value at $\boldsymbol{\pi}$. (3) For any $a \neq 0$ in F_∞ the leading coefficient $b'_N(a)$ of $\varphi'_a = \sum_{i=N}^{-\infty} b'_i(a)\tau^i$ has index $N = rv_p(a)$. In fact, all homomorphisms $\varphi'' : F_\infty \to \overline{\mathbb{F}}_p((\tau^{-1}))$ such that the dominant term in φ''_a has degree $|a|^r = p^{rv_p(a)}$ for all a in A, are conjugate by an inner automorphism of $\overline{\mathbb{F}}_p((\tau^{-1}))$; φ' can be chosen to be any of these φ''.

Lemma 5.3. *The centralizer D_∞ of the image $\mathbb{F}_q((\tau^{rv_p(\boldsymbol{\pi})}))$ of φ' in $\overline{\mathbb{F}}_p((\tau^{-1}))$ is a division algebra of rank r central over F_∞.*

Proof. The centralizer D_∞ is equal to $\mathbb{F}_{q^r}((\tau^{v_p(\boldsymbol{\pi})}))$. It is a division algebra, whose center is the field $\mathbb{F}_q((\tau^{rv_p(\boldsymbol{\pi})})) = \mathbb{F}_q((\varphi'_{\boldsymbol{\pi}})) \simeq F_\infty$, as asserted. $\square$

Remark 5.2. The invariant of the division algebra D_∞ over F_∞ is $-1/r$.

Let $\mathbb{B}$ be a perfect A-algebra and $\varphi : F_\infty \to \mathbb{B}((\tau^{-1}))$ the homomorphism of Lemma 5.1.

Definition 5.3. Let G_φ be the functor from the category of perfect $\mathbb{B}$-algebras C to the category of sets, which associates to C the set of pairs (λ, u). Here $\lambda : \mathbb{F}_{q^r} \to C$ is a homomorphism, which defines a homomorphism $\lambda_* : \mathbb{F}_{q^r}((\tau^{-1})) \to C((\tau^{-1}))$ by its action on the coefficients, and u is an element in $C[[\tau^{-1}]]$ with an invertible constant term, such that $u\varphi_a = \lambda_*(\varphi'_a)u$ for all a in F_∞.

Let Fr denote the Frobenius endomorphism $x \mapsto x^p$ of $\overline{\mathbb{F}}_p$, in particular of $\mathbb{F}_{q^r}$. Let det : $D_\infty^\times \to F_\infty^\times$ denote the reduced norm on $D_\infty^\times$, and put $\widetilde{v}_p(g) = \log_p |\det g|$ for g in $D_\infty^\times$. Then $\widetilde{v}_p(g) = v_p(\det g)$, and $\widetilde{v}_p(\tau^{v_p(\boldsymbol{\pi})}) = \widetilde{v}_p(\boldsymbol{\pi}) = v_p(\boldsymbol{\pi}^r) = rv_p(\boldsymbol{\pi})$.

Lemma 5.4. *The multiplicative group $D_\infty^\times$ of the division algebra D_∞ acts on the functor G_φ as follows: g in $D_\infty^\times$ maps (λ, u) in $G_\varphi(C)$ to*

$$g(\lambda, u) = (\lambda \circ \mathrm{Fr}^n, \tau^n \lambda_*(g)u),$$

where $n = -\widetilde{v}_p(g)$. The element $\tau^{rv_p(\boldsymbol{\pi})}$ of $D_\infty^\times$ acts trivially.

Proof. We have to check that $g(\lambda, u)$ lies in $G_\varphi(C)$. Since

$$\tau^n \varphi'_a = \mathrm{Fr}^n \cdot \varphi'_a \tau^n$$

and g lies in the centralizer D_∞ of $\varphi'(F_\infty)$, we have

$$(\lambda \circ \mathrm{Fr}^n)_* \varphi'_a [\tau^n \lambda_*(g)u] = \lambda_*(\tau^n \varphi'_a)\lambda_*(g)u = \tau^n \lambda_*(g)\lambda_*(\varphi'_a)u = [\tau^n \lambda_*(g)u]\varphi_a$$

for all a in F_∞, as required. It is clear that $D_\infty^\times$ acts on G_φ, namely $(gh)(\lambda, u) = g(h(\lambda, u))$. Since $\tau^{rv_p(\boldsymbol{\pi})}$ acts trivially on $\mathbb{F}_{q^r}$, it acts trivially on G_φ. $\square$

Remark 5.3.   (1) Lemma 5.4 establishes an action of $D_\infty^\times/\langle\pi\rangle$ on G_φ.

(2) The element $\Phi = \tau^{\log_p q}$ of $D_\infty^\times$ satisfies $\widetilde{v}_p(\Phi) = \log_p q$, and it acts by $\Phi(\lambda, u) = (\lambda \cdot \mathrm{Fr}^{-\log_p q}, u)$. Let D_0 be the kernel of the map $\widetilde{v}_p :$ $D_\infty^\times \to \mathbb{Z}$. Then g in D_0 acts on $G_\varphi(C)$ by $g(\lambda, u) = (\lambda, \lambda_*(g)u)$. The group $D_\infty^\times$ is generated by Φ over D_0.

Let $\psi : A \to \mathbb{B}[\tau]$ be another elliptic module of rank r over $\mathbb{B}$. Let $h = \sum_{i=0}^m b_i \tau^i$ in $\mathbb{B}[\tau]$ be an isogeny from φ to ψ of degree m; thus $h\varphi_a = \psi_a h$ for all a in A, and the leading coefficient b_m of h is invertible. Hence h is invertible in $\mathbb{B}((\tau^{-1}))$.

Lemma 5.5. *The morphism* $h_* : G_\varphi \to G_\psi$, *which, for each perfect* $\mathbb{B}$-*algebra* C, *maps the element* (λ, u) *in* $G_\varphi(C)$ *to* $h_*(\lambda, u) = (\lambda \circ \mathrm{Fr}^m, \tau^m uh^{-1})$ *in* $G_\psi(C)$, *is an isomorphism of functors. Moreover,* h_* *commutes with the action of* $D_\infty^\times$.

Proof. To verify the last claim, note that $\tau^m \lambda_*(g) = (\lambda \circ \mathrm{Fr}^m)_*(g)\tau^m$ for all g in $D_\infty^\times$.   □

Definition 5.4. Let j be a positive integer. The congruence subgroup D_j of $D_\infty^\times$ consists of all elements of the form $1 + \sum_{k \geq j} b_k \tau^{-kv_p(\pi)}$ in $D_\infty = \mathbb{F}_{q^r}((\tau^{v_p(\pi)}))$.

Remark 5.4. Each congruence subgroup is normal, compact, and open in $D_\infty^\times$. It has a finite index in $D_\infty^\times/\langle\pi\rangle$.

Theorem 5.6.   (*i*) *The functor* G_φ *is representable by a (perfect) ring* $\widetilde{\mathbb{B}}$ *over* $\mathbb{B}$.

(*ii*) *The stabilizer in* $D_\infty^\times$ *of any element of* $\widetilde{\mathbb{B}}$ *contains a congruence subgroup.*

(*iii*) *Let* U_∞ *be a congruence subgroup and* $\widetilde{\mathbb{B}}^{U_\infty}$ *the subring of* $\widetilde{\mathbb{B}}$ *fixed by the action of* U_∞. *Then* $\mathrm{Spec}\,\widetilde{\mathbb{B}}^{U_\infty}$ *is a finite flat étale Galois covering of* $\mathrm{Spec}\,\mathbb{B}$ *with Galois group* $D_\infty^\times/U_\infty\langle\pi\rangle$.

Proof.   (i) Let $\widetilde{\mathbb{B}}$ be the quotient of the free ring $\mathbb{B}[y, c_i(i \geq -1)]$ by the ideal generated by the following relations. The first relation is (1) $c_0 c_{-1} = 1$. To state the second, let $P(x)$ be an irreducible polynomial over $\mathbb{F}_q$ whose splitting field is $\mathbb{F}_{q^r}$. The second relation is (2) $P(y) = 0$. For the third relation, note that we have a map $\lambda : \mathbb{F}_{q^r} \to \mathbb{B}[y]$, sending a generator of $\mathbb{F}_{q^r}$ over $\mathbb{F}_q$ to y. Put $u = \sum_{i=0}^\infty c_i \tau^{-i}$. The remaining relations are obtained on equating, for each i in $\mathbb{Z}$, the coefficients of τ^i in each of the identities (3) $u\varphi_a = \lambda_*(\varphi_a')u$, for each a in F_∞. It is clear that $\widetilde{\mathbb{B}}$ represents G_φ, once we show the following.

Lemma 5.7. *The ring* $\widetilde{\mathbb{B}}$ *is a perfect ring.*

Proof. It suffices to show that $\widetilde{\mathbb{B}}$ is reduced and that each of the generators c_i is a pth power in $\widetilde{\mathbb{B}}$. We shall use the identity $u\varphi_a = \lambda_*(\varphi_a')u$, with $a = \pi^{-1}$.

Then $\varphi'_{\pi^{-1}} = \tau^{r \log_p q}$, and $\varphi_\pi = \sum_{j \geq 0} b_j(\pi) \tau^{-j-r \log_p q}$. Equating the coefficients of τ^{-i} in

$$\sum_{i \geq 0} c_i \tau^{-i} = u = \tau^{r \log_p q} u \varphi_\pi = \tau^{r \log_p q} \sum_{i \geq 0} c_i \tau^{-i} \sum_{j \geq 0} b_j(\pi) \tau^{-j-r \log_p q}$$

$$= \sum_{i,j \geq 0} c_i^{q^r} b_j(\pi)^{q^r/p^i} \tau^{-(i+j)};$$

we conclude that each c_i is a pth power in $\widetilde{\mathbb{B}}$, since

$$c_i = \left(\sum_{j+k=i} c_k b_j(\pi)^{p^{-i}} \right)^{q^r}.$$

But these are the only relations among the c_i. Hence $\widetilde{\mathbb{B}}$ is reduced and the lemma follows. □

(ii) The group $D_\infty^\times$ acts on the functor G_φ, hence on the perfect ring $\widetilde{\mathbb{B}}$. The normal subgroup D_0 acts trivially on y. If $j > 0$ then the element $1 + \tau^{-j}$ of $D_\infty^\times$ has $\tilde{v}_p(1 + \tau^{-j}) = 0$. It acts by

$$u = \sum_{i \geq 0} c_i \tau^{-i} \mapsto (1 + \tau^{-j}) u = \sum_{0 \leq i < j} c_i \tau^{-i} + \sum_{i \geq j} (c_i + c_{i-j}^{p^{-j}}) \tau^{-i}.$$

Namely $1 + \tau^{-j}$ maps c_i to c_i if $0 \leq i < j$, and to $c_i + c_{i-j}^{p^{-j}}$ if $i \geq j$. Similarly we have that each element of the congruence subgroup D_j acts trivially on c_i for i in $0 \leq i < j$. Now every element b of $\widetilde{\mathbb{B}}$ is a polynomial in only finitely many c_i's, say $0 \leq i < j$, over $\mathbb{B}[y]$, and it is stabilized by D_j.

(iii) Denote by $\mathbb{B}_j$ the image of $\mathbb{B}[y, c_i(i < j)]$ in $\widetilde{\mathbb{B}}$, where $j \geq 0$. If $U_\infty = D_j$ then $\widetilde{\mathbb{B}}^{U_\infty}$ is equal to $\mathbb{B}_j$. The coefficient $b_0(\pi)$ of φ_π, defined in the proof of Lemma 5.7, is a unit in $\mathbb{B}$. Hence the last displayed formula (for c_i) in the proof of Lemma 5.7 implies that $\mathbb{B}_{j+1}$ is a free $\mathbb{B}_j$-module ($j \geq 0$), and $\mathbb{B}_0$ is a free $\mathbb{B}[y]/(P(y))$-module, of finite rank (bounded by q^r in all cases). Consequently $\mathbb{B}_j$ is a finite flat $\mathbb{B}$-module.

To show that $\mathbb{B}_j$ is étale over $\mathbb{B}$ we shall now show that $\mathbb{B}_i$ is étale over $\mathbb{B}_{i-1}$ for all $i \geq 1$ (and $\mathbb{B}_0$ over $\mathbb{B}[y]/(P(y))$). Thus let K be a separably closed field over $\mathbb{B}_{i-1}$. We have a homomorphism $h : \mathbb{B}_{i-1} \to K$. Applying h to the last displayed formula (for c_i) in the proof of Lemma 5.7 we obtain a separable equation for c_i over K. Hence $\mathbb{B}_i \otimes_{\mathbb{B}_{i-1}} K$ is a direct product of copies of K, and $\mathbb{B}_i$ is indeed étale over $\mathbb{B}_{i-1}$, as required. Since $D_\infty^\times / U_\infty \langle \pi \rangle$ acts on $\mathbb{B}_j = \widetilde{\mathbb{B}}^{U_\infty}$ without fixed points, $\widetilde{\mathbb{B}}^{U_\infty}$ is a finite étale Galois covering of $\mathbb{B}$, with Galois group $D_\infty^\times / U_\infty \langle \pi \rangle$, and the theorem follows. □

To formulate the following proposition we recall

Definition 5.5. A ring homomorphism $D \to \mathbb{D}$ is called *radical* if for each algebraically closed field K the homomorphism $\mathrm{Hom}(\mathbb{D}, K) \to \mathrm{Hom}(D, K)$ is injective. Equivalently, $\mathrm{Spec}\,\mathbb{D} \to \mathrm{Spec}\,D$ is universally injective.

Remark 5.5. See [EGA], I. (3.7.2) and (3.7.1) for equivalent definitions.

Proposition 5.8. *Let $D \to \mathbb{D}$ be a flat radical homomorphism such that* Spec $\mathbb{D} \to$ Spec D *is surjective. Then the categories of étale covers of D and* $\mathbb{D}$ *are equivalent.*

Proof. See [SGA1], Exp. IX, Corollary 4.11, p. 241. □

We can now construct the covering scheme $\widetilde{M}_r$. Recall that $M_r =$ Spec B, where $B = A_r$ is a Noetherian ring over A. Denote by (φ, ψ) the universal elliptic module of rank r with its level structure. Let $\mathbb{B} = \varinjlim_n B^{p^{-n}}$ be the perfect closure of B. Theorem 5.6 constructs a B-algebra $\widetilde{\mathbb{B}}$ with an action of $D_\infty^\times/\langle \pi \rangle$. For any congruence subgroup U_∞ in $D_\infty^\times$, the $\mathbb{B}$-subalgebra $\widetilde{\mathbb{B}}^{U_\infty}$ of $\widetilde{\mathbb{B}}$ stabilized by U_∞ is étale over $\mathbb{B}$ with Galois group $D_\infty^\times/U_\infty\langle \pi \rangle$. It is easy to check that the homomorphism $B \to \mathbb{B}$ is radical and flat. The morphism Spec $\mathbb{B} \to$ Spec B is onto since the radical $\mathbf{m} = \varinjlim_n m^{p^{-n}}$ of the prime ideal m in Spec B is a prime ideal of $\mathbb{B}$ with $\mathbf{m} \cap B = m$. Hence we can make the

Definition 5.6. (i) Let $\widetilde{B}^{U_\infty}$ be the étale B-algebra which corresponds to $\widetilde{\mathbb{B}}^{U_\infty}$ by Prop. 5.8 and the homomorphism $B \to \mathbb{B}$. (ii) Put $\widetilde{M}_{r,U_\infty} =$ Spec $\widetilde{B}^{U_\infty}$, $\widetilde{B} = \varinjlim \widetilde{B}^{U_\infty}$, and $\widetilde{M}_r =$ Spec $\widetilde{B} = \varprojlim \widetilde{M}_{r,U_\infty}$.

Remark 5.6. (1) The ring $\widetilde{B}^{U_\infty}$ is an étale cover of B with Galois group $D_\infty^\times/U_\infty\langle \pi \rangle$.

(2) We have $\widetilde{M}_{r,U_\infty} = U_\infty \backslash \widetilde{M}_r$. In Chaps. 6 and 10 we need to use only $\widetilde{M}_{r,U_\infty}$, and not $\widetilde{M}_r$.

The group $F^\times$ embeds diagonally in $(U_\infty \backslash D_\infty^\times) \times \mathrm{GL}(r, \mathbb{A}_f)$.

Proposition 5.9. *The actions of Lemma 5.4 and Definition 5.6 define an action of $((U_\infty \backslash D_\infty^\times) \times \mathrm{GL}(r, \mathbb{A}_f))/F^\times$ on $U_\infty \backslash \widetilde{M}_r$ for each normal open subgroup U_∞ of $D_\infty^\times$ of finite index, equivalently, of $(D_\infty^\times \times \mathrm{GL}(r, \mathbb{A}_f))/F^\times$ on $\widetilde{M}_r$.*

Proof. (i) The action of $D_\infty^\times$ on the functor G_φ in Lemma 5.4 defines an action of $D_\infty^\times$ on the B-algebra $\widetilde{\mathbb{B}}$ and B-algebra $\widetilde{B}$, hence an action on $\widetilde{M}_r$. On the other hand, $\mathrm{GL}(r, \mathbb{A}_f) \cap M(r, \widehat{A})$ acts on the functor $\mathbb{F}_r$ which is represented by the scheme $M_r =$ Spec B. As in Prop. 4.15, g in $\mathrm{GL}(r, \mathbb{A}_f) \cap M(r, \widehat{A})$ defines an isogeny from the universal elliptic module φ over B to $g\varphi$. This defines (by Theorem 5.6) an action g_* on the functor G_φ, which commutes—by Lemma 5.5—with the action of $D_\infty^\times$ on M_r. In particular g lifts to an endomorphism of $\widetilde{M}_r$.

(ii) For each $a \neq 0$ in A, the elements (1) a of $\mathrm{GL}(r, \mathbb{A}_f)$ and (2) a^{-1} of $D_\infty^\times$ act on $\widetilde{M}_r$ in the same way. Indeed, (1) if $s = \varphi_a$, then $m = rv_p(a)$ and $s_*(\lambda, u) = (\lambda \circ \mathrm{Fr}^{rv_p(a)}, \tau^{rv_p(a)} u \varphi_a^{-1})$; (2) via $A - \{0\} \hookrightarrow F^\times \hookrightarrow F_\infty^\times \hookrightarrow D_\infty^\times$, a maps to φ_a'. Since $|\det \varphi_a'| = |a|_\infty^r$, we have $n = -rv_p(a)$, and $a(\lambda, u) = (\lambda \circ \mathrm{Fr}^{-rv_p(a)}, \tau^{-rv_p(a)} \lambda_*(\varphi_a') u)$. Since $\lambda_*(\varphi_a') u \varphi_a^{-1} = u$, the composition of these two maps is the identity.

(iii) We obtained an action of $[\mathrm{GL}(r, \mathbb{A}_f) \cap M(r, \widehat{A})] \times D_\infty^\times$, such that for each $a \neq 0$ in A, the diagonal element (a, a) acts trivially. Hence we obtain an action of $\mathrm{GL}(r, \mathbb{A}_f) \times D_\infty^\times$, where $F^\times$ acts trivially, as required.

$\square$

Part 2. Hecke Correspondences

In this part and in Parts 3 and 4 we study some relations between two natural actions—of a Galois group and of a Hecke algebra—on ℓ-adic cohomology groups with compact support and coefficients in a $\overline{\mathbb{Q}}_\ell$-sheaf $\mathbb{L}(\rho)$, attached to the geometric generic fiber $M_{r,I} \otimes_A \overline{F}$ of the moduli scheme $M_{r,I}$ constructed in Part 1, Chap. 4. We present two approaches. That of Part 2, Sect. 6.2, is based on congruence relations and Hecke correspondences. That of Part 4 is an application of the trace formula. Chapters 7–9 of Part 3 develop the tools needed for the comparison in Chaps. 10–11 of Part 4 of the Grothendieck–Lefschetz fixed point formula with the Selberg trace formula. We begin with a summary (in Sect. 6.1) of those properties of the ℓ-adic cohomology groups with compact support which we need in order to state—and use in the proof of—our main theorems in Chaps. 6, 10, and 11. The main application of the theory of congruence relations is given in Theorem 6.10, which asserts that if $\pi_f \times \sigma$ is an irreducible $G(\mathbb{A}_f) \times \mathrm{Gal}(\overline{F}/F)$-subquotient of $H_c^i(M_{r,I} \times_A \overline{F}, \mathbb{L}(\rho))$, then for almost all v each eigenvalue of the geometric Frobenius endomorphism $\sigma(\mathrm{Fr}_v \times 1)$ at v is the product by $q_v^{(r-1)/2}$ of a Hecke eigenvalue of the component π_v of π at v. Consequently $\sigma(\mathrm{Fr}_v \times 1)$ has at most r distinct eigenvalues.

6. DELIGNE'S CONJECTURE AND CONGRUENCE RELATIONS

6.1. ℓ-Adic Cohomology. In this Sect. 6.1 we summarize some properties of the étale cohomology groups with compact support needed for our study of the action of the Hecke operators and the Galois group on them. This is a rather selective summary, and not a complete exposition. For an introductory textbook to the subject see [Mi]. The shorter exposition of [SGA4 1/2], Arcata, Rapport, is very useful, and so are the fundamental results of [SGA4], Exp. XVII, XVIII, and [SGA5], Exp. III.

6.1.1. Throughout this section, by a scheme we mean a separated scheme of finite type over a scheme S. Basic definitions now follow. An *étale covering* of a scheme U is a finite set $g_i : Z_i \to U$ of étale morphisms with $U = \cup_i g_i(Z_i)$. The (small) *étale site* $X_{\text{ét}}$ of X is the category whose objects are all the étale morphisms $f : U \to X$, and whose morphisms are the étale morphisms $U \to U'$ over X, together with the étale topology, which is defined to be the Grothendieck topology (see [SGA4 1/2], p. 15), in which the coverings of U are the étale coverings. A *sheaf* (resp. of sets) on the site $X_{\text{ét}}$ is a contravariant functor from the category underlying $X_{\text{ét}}$ to the category of abelian groups (resp. sets) which satisfies standard axioms (see [Mi], p. 49). The sheaves on $X_{\text{ét}}$ make a category (see [Mi], p. 50).

The simplest example of a sheaf (resp. of sets) on $X_{\text{ét}}$ is the *constant sheaf* N_X associated to a finite abelian group (resp. finite set) N. The constant sheaf N_X assigns to each étale morphism $g : Z \to X$ the abelian group (resp. set) $\text{Hom}(Z, N) = N^{\pi_0(Z)}$; here $\pi_0(Z)$ is the finite set of connected components of Z. Denote by $\text{Hom}_X(Z, N \times X)$ the group (resp. set) of morphisms $Z \to N \times X$ whose composition with the projection $N \times X \to X$ on the second factor is the morphism $g : Z \to X$. Then $\text{Hom}(Z, N) = \text{Hom}_X(Z, N \times X)$. This suggests the following definition of a locally constant sheaf.

6.1.2. Let $Y \to X$ be a finite étale morphism. Then $Z \to Y_X(Z) = \text{Hom}_X(Z, Y)$ defines a sheaf Y_X of sets on the étale site $X_{\text{ét}}$. The sheaf of set Y_X is a sheaf if Y is an abelian group scheme. A *locally constant* sheaf on $X_{\text{ét}}$ is a sheaf which is locally constant, namely it is of the form Y_X. A sheaf L on $X_{\text{ét}}$ is locally constant if and only if (see [Mi], p. 155) there is an étale covering $Z_i \to X$ of X such that the restriction $L|Z_i$ of L to the étale site $Z_{i,\text{ét}}$ of Z_i is constant for all i. If $Y \to X$ and $Y' \to X$ are finite and étale then we have $\text{Hom}_X(Y, Y') = \text{Hom}(Y_X, Y'_X)$; the later Hom is taken in the category of sheaves on $X_{\text{ét}}$.

Remark 6.1. Our definition of "locally constant sheaf" is the same as that of a "locally constant sheaf with finite stalks" in [Mi], p. 155.

Example 6.1. Let K be a field of characteristic $p > 0$. Let n be a positive integer with $(p, n) = 1$. Let $X = Y$ be the affine group scheme $\mathbb{G}_{m,K} = \text{Spec } K[x, x^{-1}]$. Then the morphism $Y \to X$ defined by $x \mapsto x^n$ is étale, since $(n, p) = 1$, and we obtain a locally constant sheaf Y_X on $X_{\text{ét}}$. Let N be the group of nth roots of unity in the multiplicative group $K^\times$ of K. We claim that the restriction of Y_X to the étale covering $\{Z \to X\} = \{Y \to X\}$ of X is the constant sheaf defined by N. Indeed,

$$Z \times_X Y = Y \times_X Y = \{(z, y); z^n = y^n\} = \{(z, \zeta); z \in Z, \zeta \in N\} = Z \times N$$

defines a constant sheaf on $Z_{\text{ét}}$.

6.1.3. Let X be a scheme, $S(X)$ the category of sheaves on $X_{\text{ét}}$, and $f : X' \to X$ a morphism of schemes. For any sheaf L in $S(X')$ define the *direct image* $f_* L$ of L by $(f_* L)(Z) = L(Z \times_X X')$ for any Z in $X_{\text{ét}}$. Then $f_* L$ is a sheaf in $S(X)$ (see [Mi], p. 59). Moreover, $f_* : S(X') \to S(X)$ is a functor which is exact if f is a finite morphism, e.g., a closed immersion (see [Mi], p. 72). If L is the locally constant sheaf $Y_{X'}$ defined by the finite étale morphism $Y \to X'$, and $f : X' \to X$ is finite étale, then the direct image $f_* Y_{X'}$ is locally constant, defined by the finite étale morphism $Y \to X' \to X$. Indeed,

$$(f_* Y_{X'})(Z) = \text{Hom}_{X'}(Z \times_X X', Y) = \text{Hom}_X(Z, Y) = Y_X(Z).$$

The direct image functor $f_* : S(X') \to S(X)$ has a left adjoint functor (see [Mi], p. 68) which is denoted by $f^* : S(X) \to S(X')$ and called the *inverse image* functor. Thus f^* is the unique functor which satisfies

$$\text{Hom}(f^* L, L') = \text{Hom}(L, f_* L') \quad (L \in S(X), L' \in S(X')).$$

If Y_X is a locally constant sheaf on X, then its inverse image $f^* Y_X$ is the locally constant sheaf $Y'_{X'}$ on X' defined by the finite étale morphism $Y' = Y \times_X X' \to X'$; indeed, $\text{Hom}_{X'}(Z', Y') = \text{Hom}_X(Z', Y)$. If f is finite and étale then the inverse image $f^* Y_X$ is the restriction of the sheaf Y_X from the étale site $X_{\text{ét}}$ to $X'_{\text{ét}}$. Hence for all L in $S(X)$ we define the restriction $L|X'$ of L to X' to be $f^* L$.

The direct image of a locally constant sheaf is not necessarily locally constant when f is not finite étale. A sheaf L on $X_{\text{ét}}$ is called *constructible* (see [Mi], p. 161) if every irreducible Zariski closed subscheme Z of X contains a nonempty open subscheme U such that the restriction $L|U$ of L to the étale site of U is locally constant. Equivalently L is called constructible if X is the union of locally closed subschemes U_i such that each $L|U_i$ is locally constant. If f is a proper (in particular a finite) morphism, and L' is a constructible (in particular locally constant) sheaf on X', then the sheaf $f_* L'$ is constructible (see [Mi], p. 223).

6.1.4. Fix a prime $\ell \neq p$. A $\mathbb{Z}_\ell$-*sheaf* $\mathbb{L}$ on X is defined to be a projective system of constructible sheaves L_n of $\mathbb{Z}/\ell^n \mathbb{Z}$-modules such that the transition morphisms $L_n \to L_{n-1}$ factorize via the isomorphism

$$L_n \otimes_{\mathbb{Z}/\ell^n \mathbb{Z}} \mathbb{Z}/\ell^{n-1} \mathbb{Z} \overset{\sim}{\to} L_{n-1} \quad \text{for all } n;$$

see [Mi], p. 163. The $\mathbb{Z}_\ell$-sheaf $\mathbb{L}$ is called *smooth* if the L_n are locally constant. We also introduce the category of (smooth) $\mathbb{Q}_\ell$-*sheaves* to be the category whose objects are the (smooth) $\mathbb{Z}_\ell$-sheaves and whose morphisms are given by

$$\mathrm{Hom}(\mathbb{L} \otimes \mathbb{Q}_\ell, \mathbb{L}' \otimes \mathbb{Q}_\ell) = \mathrm{Hom}(\mathbb{L}, \mathbb{L}') \otimes \mathbb{Q}_\ell.$$

Here $\mathbb{L} \otimes \mathbb{Q}_\ell$ denotes the $\mathbb{Z}_\ell$-sheaf $\mathbb{L}$ viewed as a $\mathbb{Q}_\ell$-sheaf. If no confusion is likely to occur we write "$\mathbb{Q}_\ell$-sheaf $\mathbb{L}$" for $\mathbb{L} \otimes \mathbb{Q}_\ell$. The *fiber* of $\mathbb{L} \otimes \mathbb{Q}_\ell$ at a geometric point $\overline{x}$ of X is defined to be

$$(\mathbb{L} \otimes \mathbb{Q}_\ell)_{\overline{x}} = \mathbb{L}_{\overline{x}} \otimes_{\mathbb{Z}_\ell} \mathbb{Q}_\ell.$$

Example 6.2. The scheme $Y_n = \mathbb{G}_{m,K}$ with the morphism $p_n : Y_n \to X = \mathbb{G}_{m,K}$ given by $p_n(x) = x^{\ell^n}$ defines a locally constant sheaf L_n which is locally (in the étale topology) isomorphic to $\mathbb{Z}/\ell^n\mathbb{Z}$ (since $Y_n \times_X Y_n = Y_n \times \mathbb{Z}/\ell^n\mathbb{Z}$). The transition morphisms $\ell : Y_n \to Y_{n-1}$, $x \mapsto x^\ell$, define a smooth sheaf $\mathbb{L}$.

In this work we shall be interested mainly in the following fundamental:

Example 6.3. Let X be a scheme, $\widetilde{X}$ a finite étale Galois covering of X with Galois group $\pi_1 = \pi_1(\widetilde{X}/X)$, and $\rho : \pi_1 \to \mathrm{Aut}\,\mathbb{Q}_\ell^t$ an irreducible (finite dimensional) representation. Since π_1 is finite, in particular compact, ρ factorizes through $\rho' : \pi_1 \to \mathrm{Aut}\,\mathbb{Z}_\ell^t$. Then π_1 acts on $\widetilde{X} \times (\mathbb{Z}/\ell^n\mathbb{Z})^t$ by $g(\widetilde{x}, v) = (g\widetilde{x}, \rho'(g)v)$. The quotient Y_n has the property that the natural projection $p_n : Y_n \to X$ is finite and étale. Locally, in the étale topology, Y_n is a trivial bundle with fiber $(\mathbb{Z}/\ell^n\mathbb{Z})^t$. The projective system $\mathbb{L}(\rho)$, of locally constant sheaves $L_n = Y_{n,X}$ of $\mathbb{Z}/\ell^n\mathbb{Z}$-modules defined by the morphisms $p_n : Y_n \to X$, is a smooth $\mathbb{Z}_\ell$-sheaf.

Suppose that G is a topological group with the property that the direct product $G \times \pi_1$ acts on $\widetilde{X}$. Then G acts on X, and we assume (in our example) that G acts on X without fixed points. Fix n, put Y for Y_n and L for the locally constant sheaf determined by Y. Let U be an open subgroup of G. Put X_U for $U\backslash X$. Put L_U for the locally constant sheaf on (the étale site of) X_U defined by $Y_U = U\backslash Y(= \pi_1\backslash[(U\backslash\widetilde{X}) \times (\mathbb{Z}/\ell^n\mathbb{Z})^t])$. Let g be an element of G. Then there are morphisms

$$h : U \cap g^{-1}Ug\backslash X \to U\backslash X \quad \text{(quotient by } U\text{)},$$

$$f : U \cap gUg^{-1}\backslash X \to U\backslash X \quad \text{(quotient by } U\text{)},$$

$$g : U \cap g^{-1}Ug\backslash Y \to U \cap gUg^{-1}\backslash Y \quad \text{(multiplication by } g\text{)}.$$

The sheaf $L_{U\cap g^{-1}Ug}$ on $U \cap g^{-1}Ug\backslash X$ determined by $U \cap g^{-1}Ug\backslash Y$ is isomorphic to h^*L_U, since the morphism

$$U\backslash Y \to (U\backslash X) \times_{(U\cap g^{-1}Ug\backslash X)} (U \cap g^{-1}Ug\backslash Y)$$

is an isomorphism. Similarly we have $f^*L_U = L_{U\cap gUg^{-1}}$. The morphism g defines a sheaf morphism $L_{U\cap g^{-1}Ug} \to L_{U\cap gUg^{-1}}$. Hence a sheaf morphism $\beta = \beta(g) : h^*L_U \to f^*L_U$. Clearly f^*L_U can be viewed as a locally constant sheaf on $U \cap g^{-1}Ug\backslash X$ defined by the morphism

$$U \cap gUg^{-1}\backslash Y \to (U \cap gUg^{-1}\backslash X \xrightarrow{g^{-1}}) U \cap g^{-1}Ug\backslash X.$$

We shall return below to the example above. Now we make the following:

Definition 6.1. (a) If L is a constructible sheaf (of finite abelian groups, in particular a locally constant sheaf Y_X defined by an abelian group scheme Y which is finite and étale over X), denote by $H_c^i(X, L)$ the *étale cohomology group* with compact support and coefficients in the sheaf L of the scheme X (see [Mi], p. 227). When X is a separated scheme of finite type over a separably closed field (this is the only case used below), these are finite groups, defined for $i \geq 0$, with $H_c^i(X, L) = 0$ for $i > 2 \dim X$ (see [Mi], p. 221). Note that $H_c^i(\overline{X}, L)$ is defined to be $H^i(\overline{X}, j_! L)$ for any compactification $j : X \hookrightarrow \overline{X}$ of X, where $j_! L$ is the "extension by zero" of L to $\overline{X}$; it is independent of the choice of compactification.

(b) The *ℓ-adic cohomology group* $H_c^i(X, \mathbb{L})$ of X with compact support and coefficients in the $\mathbb{Z}_\ell$-sheaf $\mathbb{L}$ is the projective limit $\varprojlim_n H_c^i(X, L_n)$. It is a $\mathbb{Z}_\ell$-module of finite rank, which vanishes unless $0 \leq i \leq 2 \dim X$. For a $\mathbb{Q}_\ell$-sheaf $\mathbb{L} = \mathbb{L}_0 \otimes \mathbb{Q}_\ell$ we write $H_c^i(X, \mathbb{L}) = H_c^i(X, \mathbb{L}_0) \otimes_{\mathbb{Z}_\ell} \mathbb{Q}_\ell$. The effect of tensoring with the field $\mathbb{Q}_\ell$ is to make $H_c^i(X, \mathbb{L})$ into a vector space, and so erase the torsion in the $\mathbb{Z}_\ell$-module $H_c^i(X, \mathbb{L}_0)$.

Remark 6.2. Let E_λ be a finite field extension of $\mathbb{Q}_\ell$, R_λ the ring of integers in E_λ, and π a uniformizer in R_λ. Then we can define an R_λ-sheaf as above on replacing $\mathbb{Z}/\ell^n\mathbb{Z}$ by $R_\lambda/\pi^n R_\lambda$. An R_λ-sheaf is equivalent to a $\mathbb{Z}_\ell$-sheaf $\mathbb{L}$ together with a homomorphism $R_\lambda \to \text{End } \mathbb{L}$. Each λ-adic cohomology group is an R_λ-module of finite rank; it is constructed as a $\mathbb{Z}_\ell$-module on which R_λ acts (see [SGA4 1/2], p. 85). Similarly we have E_λ-sheaves and also $\overline{\mathbb{Q}}_\ell$-sheaves, where $\overline{\mathbb{Q}}_\ell$ is an algebraic closure of $\mathbb{Q}_\ell$.

Let $f : X' \to X$ be a finite morphism. The inverse image functor f^* yields, for each constructible sheaf L on X and $\mathbb{Z}_\ell$-sheaf $\mathbb{L}$ on X, the inverse image homomorphisms

$$H_c^i f^* : H_c^i(X, L) \to H_c^i(X', f^* L)$$

and

$$H_c^i f^* : H_c^i(X, \mathbb{L}) \to H_c^i(X', f^* \mathbb{L});$$

$f^* \mathbb{L}$ is the $\mathbb{Z}_\ell$-sheaf obtained as the inverse image of the constructible sheaf $\mathbb{L}$ on X. When f is finite, the direct image functor f_* is exact (see [Mi], p. 72, and [SGA4 1/2], p. 24), and we have the direct image isomorphisms

$$H_c^i f_* : H_c^i(X', L') \xrightarrow{\sim} H_c^i(X, f_* L')$$

and

$$H_c^i f_* : H_c^i(X', \mathbb{L}') \xrightarrow{\sim} H_c^i(X, f_* \mathbb{L}')$$

for any constructible sheaf L' on X' and $\mathbb{Z}_\ell$-sheaf $\mathbb{L}'$ on X'. Here $f_* \mathbb{L}'$ is the $\mathbb{Z}_\ell$-sheaf obtained as the direct image of the $\mathbb{Z}_\ell$-sheaf $\mathbb{L}'$ on X'.

6.1.5. If $f : X' \to X$ is finite and étale then the functor f^* is right adjoint to f_*, namely $\mathrm{Hom}(f_* L', L) = \mathrm{Hom}(L', f^* L)$ for all sheaves L', L on X', X. For a proper morphism f the functor Rf_* ($= Rf_!$) has a right adjoint, denoted $Rf^!$, only in the derived category $D(X)$ (see [SGA4 1/2], C, D and [Mi], p. 310) of the category $S(X)$ of sheaves on X. However, when f is finite, which is the case of interest for us, the functor f_* is exact ([Mi], p. 72), in particular left exact. Hence it has a right adjoint functor $f^! : S(X) \to S(X')$. Thus by definition we have

$$\mathrm{Hom}(f_* L', L) = \mathrm{Hom}(L', f^! L) \qquad (L \in S(X), L' \in S(X'))$$

for a finite morphism f. In particular, if $L' = f^! L$, we obtain

$$\mathrm{Hom}(f_* f^! L, L) = \mathrm{Hom}(f^! L, f^! L).$$

Definition 6.2. (i) Let $f, h : X' \to X$ be finite morphisms, L a smooth sheaf on X, and $\alpha : h^* L \to f^! L$ a sheaf morphism. Let $H_c^i(f, \alpha, h)$ be the endomorphism of the $\mathbb{Z}_\ell$-module $H_c^i(X, L)$ defined as the composition of

$$H_c^i(X, L) \xrightarrow{H_c^i h^*} H_c^i(X', h^* L) \xrightarrow{H_c^i \alpha} H_c^i(X', f^! L) \xrightarrow{H_c^i f_*}$$
$$H_c^i(X, f_* f^! L) \xrightarrow{t} H_c^i(X, L),$$

where $t : f_* f^! L \to L$ is the sheaf morphism corresponding to the identity morphism $f^! L \to f^! L$ and $H_c^i f_*$ is an isomorphism.

(ii) Suppose that $f : X' \to X$ is a finite flat morphism. Let $a : f^* L \to f^! L$ denote the morphism obtained by adjunction from the trace map $\mathrm{tr} : f_* f^* L \to L$ of [SGA4], Exp. XVIII, Theorem 2.9 (p. 553).

(iii) In the notations of (i), given a sheaf morphism $\beta : h^* L \to f^* L$ we write $H_c^i(f, \beta, h)$ for $H_c^i(f, \alpha, h)$, where $\alpha = a \circ \beta$.

Remark 6.3. If $f = h$, and $a : f^* L \to f^! L$ is an isomorphism, and $\beta : h^* L \to f^* L$ is the identity, then $H_c^i(f, \mathrm{id}, f)$ is the multiplication by the degree of X' over X (cf. [SGA4], XVIII, p. 554).

6.1.6. Our next aim is to compare the cohomologies of geometric fibers of X and construct the endomorphisms $H_c^i(f, \alpha, h)$ in a way compatible with this fibration.

We begin by considering a morphism $b : X \to S$ of schemes and a $\mathbb{Z}_\ell$-sheaf L on X. Let $j : S_0 \to S$ be an open dense subscheme of S. For each $i \geq 0$ denote by $R^i b_! L | S_0$ the cohomology sheaf $H^i(\_, Rb_! L | S_0)$ of the restriction $Rb_! L | S_0 = j^* Rb_! L$ to S_0 of the complex $Rb_! L$ of sheaves in the derived category $D(X)$.

Lemma 6.1. *For any $\mathbb{Z}_\ell$-sheaf L on X and morphism $b : X \to S$, the sheaf $R^i b_! L$ is constructible. Consequently there is an open dense subscheme $j : S_0 \to S$ such that the restriction $R^i b_! L | S_0$ is a smooth sheaf, for all $i \geq 0$.*

Proof. The first claim is the constructibility theorem of [SGA4], XVII, p. 364. See also [SGA4 1/2], Theorem of finitude, Theorem 1.9, p. 236. The second follows from the definition of constructibility.   □

Let S be an irreducible scheme. Then any open dense subscheme S_0 of S is also irreducible. Let $\eta : \operatorname{Spec} K \to S$ be a geometric generic point of S (*geometric* means that K is a separably closed field; *generic* means that the image of η is dense in S; we use η rather than the standard $\bar{\eta}$ to simplify the notations). Then η factorizes through $S_0 \to S$. Let G be a sheaf on S. Denote by G_η the (geometric generic) stalk of G at η (see [Mi], p. 60). Let $v :$ $\operatorname{Spec} k \to S_0$ be a geometric closed (namely the image of v is closed in S_0) point. Denote by G_v the stalk of G at v. The fundamental group $\pi_1(S_0, \eta)$ of S_0 at η acts on the stalk G_η, and $\pi_1(S_0, v)$ acts on G_v. By the class of an isomorphism $i : G_\eta \stackrel{\sim}{\to} G_v$ we mean the set of isomorphisms $i' : G_\eta \stackrel{\sim}{\to} G_v$ of the form $i' = i \circ \alpha$ (α in $\pi_1(S_0, \eta)$). This is equal to the set of isomorphisms $i' = \beta \circ i$ (β in $\pi_1(S_0, v)$). With these notations, the specialization and cospecialization theorems of [SGA4 1/2], p. 256/7, assert

Lemma 6.2. *Let S be an irreducible scheme, S_0 an open dense subscheme, and G a $\mathbb{Z}_\ell$-sheaf on S whose restriction $G|S_0$ to S_0 is smooth. Then G_η is noncanonically isomorphic to G_v for any geometric closed point v in S_0, and the class of the isomorphism $G_\eta \stackrel{\sim}{\to} G_v$ is canonical.*

Let $b : X \to S$ be a morphism. Let $\mathbb{L}$ be a sheaf as in Lemma 6.1. Let $\eta :$ $\operatorname{Spec} K \to S$ be a geometric generic point. Let $v : \operatorname{Spec} k \to S_0$ be a geometric closed point in S_0 (S_0 is defined by Lemma 6.1). Let $X_\eta = X \times_S \operatorname{Spec} K$ be the (geometric generic) fiber of X at η. Let $X_v = X \times_S \operatorname{Spec} k$ be the (geometric special) fiber of X at v. We conclude

Proposition 6.3. *For any $\mathbb{Z}_\ell$-sheaf $\mathbb{L}$ on X there is an open dense subscheme S_0 of S such that $H_c^i(X_\eta, \mathbb{L}|X_\eta)$ and $H_c^i(X_v, \mathbb{L}|X_v)$ are noncanonically isomorphic for all $i \geq 0$ and any closed geometric point v in S_0. The class of this isomorphism is canonical.*

Proof. From the proper base change theorem ([SGA4], XVII, Prop. 5.2.8, p. 358) it follows that one has the canonical isomorphisms

$$(R^i b_! \mathbb{L})_s = H_c^i(X_s, \mathbb{L}|X_s) \quad \text{and} \quad (R^i b_! \mathbb{L})_v = H_c^i(X_v, \mathbb{L}|X_v).$$

The proposition now follows from Lemmas 6.1 and 6.2.   □

6.1.7.   Let $f, h : X' \to X$ be finite morphisms, and $b : X \to S$ a morphism where S is irreducible. Let $\mathbb{L}$ be a smooth sheaf on X, and fix a morphism $\alpha : h^* \mathbb{L} \to f^! \mathbb{L}$. Proposition 6.3 asserts that there is an open dense subscheme S_0 of S such that $H_c^i(X_\eta, \mathbb{L})$ and $H_c^i(X_v, \mathbb{L})$ are isomorphic for all geometric closed points v in S_0, where the class of the isomorphism is canonical. Let t be a geometric point in S_0 (in particular v or η as above). Denote by X_t and X_t' the fibers of X and X' at t, by $i_t' : X_t' \to X'$ and $i_t : X_t \to X$ the natural morphisms and by $f_t, h_t : X_t' \to X_t$ the fiber morphisms of $f, h : X' \to X$.

Definition 6.3. Let $H_c^i(f_t, \alpha, h_t)$ be the endomorphism of $H_c^i(X_t, \mathbb{L}|X_t)$ defined as the following composition:

$$H_c^i(X_t, \mathbb{L}|X_t) \xrightarrow{H_c^i h_t^*} H_c^i(X_t', h_t^*(\mathbb{L}|X_t)) \xrightarrow{H_c^i \alpha} H_c^i(X_t', f_t^!(\mathbb{L}|X_t))$$

$$\xrightarrow{H_c^i f_{t*}} H_c^i(X_t, f_{t*} f_t^!(\mathbb{L}|X_t)) \xrightarrow{t} H_c^i(X_t, \mathbb{L}|X_t).$$

The homomorphism $H_c^i \alpha$ is obtained from the sheaf morphism $\alpha : (h^* \mathbb{L})|X_t' \to (f^! \mathbb{L})|X_t'$.

The class of the $\mathbb{Z}_\ell$-module isomorphism $H_\eta \xrightarrow{\sim} H_v$ (where $H_t = H_c^i(X_t, \mathbb{L})$ with $t = \eta$ or v) is canonically determined. The functorial construction of $H_c^i(f_t, \alpha, h_t)$ implies then the following:

Proposition 6.4. *The isomorphism $H_\eta \xrightarrow{\sim} H_v$ can be chosen to map the endomorphism $H_c^i(f_\eta, \alpha, h_\eta)$ of H_η to the endomorphism $H_c^i(f_v, \alpha, h_v)$ of H_v, uniformly in f, h, and α.*

6.1.8. Let X be a scheme (as usual, X is separated of finite type over a scheme S). Let $(h, f : X' \to X)$ be a pair of finite flat morphisms $h : X' \to X$ and $f : X' \to X$. We say that the pair $(h_1, f_1 : X_1' \to X)$ is isomorphic to $(h, f : X' \to X)$ if there is an isomorphism $F : X' \to X_1'$ with $h_1 \circ F = h$ and $f_1 \circ F = f$. By a *correspondence* T on X we mean an isomorphism class of pairs. Let $R(X)$ be the quotient of the free abelian group generated (over $\mathbb{Z}$) by all correspondences on X, by the relation $T_1 = d_1 T$ if T_1 is the correspondence of $(h \circ k, f \circ k : X_1' \to X)$ and $k : X_1' \to X'$ is a finite flat morphism of degree d_1. The group $R(X)$ has a $\mathbb{Z}$-algebra structure, which is not necessarily abelian, obtained as follows. The product PP_1 of the pairs $P = (h, f : X' \to X)$ and $P_1 = (h_1, f_1 : X_1' \to X)$ is the pair $(h \circ \mathrm{pr}_1, f_1 \circ \mathrm{pr}_2 : X' \times_X X_1' \to X)$; here $\mathrm{pr}_1 : X' \times_X X_1' \to X'$ and $\mathrm{pr}_2 : X' \times_X X_1' \to X_1'$ are the natural projections. If $\mathbb{L}$ is smooth, then for each pair $(h, f : X' \to X)$ and morphism $\alpha : h^* \mathbb{L} \to f^! \mathbb{L}$ we constructed in Definition 6.2(i) an endomorphism $H_c^i(f, \alpha, h)$ of $H_c^i(X, \mathbb{L}|X)$. In particular, $H_c^i(f, \mathrm{id}, f)$ is multiplication by the degree of f.

We shall apply the above constructions in the case of Example 6.3 in Sect. 6.1.4, which is our main example in this work. As in Sect. 6.1.4, let $\tilde{X}$ be a finite étale Galois covering of a scheme X with Galois group π_1, and let G be a topological group such that $G \times \pi_1$ acts on $\tilde{X}$. For any open subgroup U of G put $X_U = U \backslash X$. Given g in G, put $X' = U \cap g^{-1} U g \backslash X$. In Sect. 6.1.4 we constructed a correspondence $h : X' \to X_U$ (quotient by U), and $f : X' \to U \cap g U g^{-1} \backslash X \to X_U$ (the first arrow is multiplication by g, the second is quotient by U; the finite morphism f can also be defined by $X' \to g^{-1} U g \backslash X \to X_U$, where the first arrow is the quotient by $g^{-1} U g$ and the second is multiplication by g), and a sheaf morphism $\beta = \beta(g) : h^* \mathbb{L}_U \to f^* \mathbb{L}_U$; here $\mathbb{L}_U$ is the smooth $\mathbb{Q}_\ell$-sheaf determined by the projective system of finite étale Galois morphisms $U \backslash Y_n \to X_U$ defined in Sect. 6.1.4. Given this data we put $\alpha(g) = a \circ \beta(g)$ and construct (as in Sect. 6.1.7) the endomorphism $H_c^i(f, \alpha(g), h)$ of $H_c^i(X_U, \mathbb{L}_U)$ as the composition of

$$H_c^i(X_U, \mathbb{L}_U) \overset{h^*}{\to} H_c^i(X', h^*\mathbb{L}_U) \overset{\alpha(g)}{\to} H_c^i(X', f^!\mathbb{L}_U)$$
$$\overset{f_*}{\to} H_c^i(X_U, f_* f^!\mathbb{L}_U) \overset{t}{\to} H_c^i(X_U, \mathbb{L}_U).$$

It is clear that $H_c^i(f, \alpha(g), h)$ depends only on the double coset $\overline{g}$ of g in $U \backslash G / U$. Hence it can be denoted by $H_c^i(\overline{g})$. Let $\mathbb{H}(U)$ be the $\mathbb{Q}_\ell$-algebra of U-double cosets in G: we fix a Haar measure on G such that the volume $|U|$ lies in $\mathbb{Q}_\ell^\times$, identify a U-double coset UgU with the quotient of its characteristic function by $|U|$, and let the product be defined by convolution. A standard verification shows that $\overline{g} \mapsto H_c^i(\overline{g})$ is an algebra homomorphism $\mathbb{H}(U) \to \operatorname{End} H_c^i(X_U, \mathbb{L}_U)$, namely that $\mathbb{H}(U)$ acts on $H_c^i(X_U, \mathbb{L}_U)$. Proposition 6.4 of Sect. 6.1.7 now implies the following:

Proposition 6.5. *Let $G, U, \widetilde{X}, \mathbb{L}$ be as above. Let $b : X \to S$ be a morphism where S is irreducible. Then there exists an open dense subscheme S_0 of S such that the $\mathbb{H}(U)$-algebras $H_c^i(X_{U,\eta}, \mathbb{L}_U)$ (as in Sect. 6.1.6, η denotes a geometric generic point in S), and $H_c^i(X_{U,v}, \mathbb{L}_U)$ are isomorphic for every geometric closed point v in S_0.*

6.1.9. Let K be a field of characteristic p. Let $\overline{K}$ be a separable closure of K. Let X be a scheme over K. Put $\overline{X} = X \otimes_K \overline{K} (= X \times_{\operatorname{Spec} K} \operatorname{Spec} \overline{K})$. Then the Galois group $\operatorname{Gal}(\overline{K}/K)$ acts continuously on $\overline{K}$, hence on $\overline{X}$. Consequently the group $H_c^i(\overline{X}, \mathbb{L})$ is a $\operatorname{Gal}(\overline{K}/K)$-module; the action of $\operatorname{Gal}(\overline{K}/K)$ is continuous.

If K is a finite field $\mathbb{F}_q$ of characteristic p, the Galois group $\operatorname{Gal}(\overline{\mathbb{F}}_q/\mathbb{F}_q)$ is topologically generated by the Frobenius substitution $\operatorname{Fr}_q : x \mapsto x^q$. The action of Fr_q on the second factor in $\overline{X} = X \otimes_{\mathbb{F}_q} \overline{\mathbb{F}}_q$ is denoted by $1 \times \operatorname{Fr}_q$ and called the *arithmetic* Frobenius. In addition Fr_q acts on X, hence on the first factor in $\overline{X} = X \otimes_{\mathbb{F}_q} \overline{\mathbb{F}}_q$. This action, denoted by $\operatorname{Fr}_q \times 1$, is called the *geometric* Frobenius. Both $\operatorname{Fr}_q \times 1$ and $1 \times \operatorname{Fr}_q$ act on $H_c^i(\overline{X}, \mathbb{L})$; their product is the identity endomorphism (see [SGA4 1/2], p. 80).

Let X be a separated scheme of finite type over $\mathbb{F}_q$, where $q = p^d$. Let $\mathbb{L} = (L_n)$ be a smooth $\mathbb{Z}_\ell$-sheaf, where L_n is a locally constant sheaf associated with the finite étale morphism $p_n : Y_n \to X$. Let m be an integer. At each point x in the set

$$X(\mathbb{F}_{q^{|m|}}) = \operatorname{Hom}(\operatorname{Spec} \mathbb{F}_{q^{|m|}}, X) = X(\overline{\mathbb{F}}_q)^{(\operatorname{Fr}_q \times 1)^m},$$

the inverse image $p_n^*(x)$ is isomorphic to $(\mathbb{Z}/\ell^n \mathbb{Z})^t$, and the stalk $\mathbb{L}_x = \varprojlim_n p_n^*(x)$ is isomorphic to $\mathbb{Z}_\ell^t$. Since $(\operatorname{Fr}_q \times 1)^m$ fixes x, it acts on $p_n^*(x)$. Passing to the limit we obtain an action of $(\operatorname{Fr}_q \times 1)^m$ on the stalk $\mathbb{L}_x \simeq \mathbb{Z}_\ell^t$. Denote by $\operatorname{tr}((\operatorname{Fr}_q \times 1)^m | \mathbb{L}_x)$ the trace of $(\operatorname{Fr}_q \times 1)^m$ on the stalk $\mathbb{L}_x$. Denote by $(\operatorname{Fr}_q \times 1)^m$ also the endomorphism of $H_c^i(\overline{X}, \mathbb{L})$ which is denoted in Definition 6.2(i) in (6.1.5) by

$$H_c^i(\operatorname{id}, (\operatorname{Fr}_q \times 1)^m, (\operatorname{Fr}_q \times 1)^m);$$

thus $f = \operatorname{id}$, $\alpha = (\operatorname{Fr}_q \times 1)^m$, and $h = (\operatorname{Fr}_q \times 1)^m$ there.

The following form of the fixed point formula, for powers of the Frobenius acting on a scheme over a finite field, is due to Grothendieck (see [SGA5], Exp. III, (6.13.3), p. 134, [SGA4 1/2], p. 86).

Theorem 6.6 (Grothendieck Fixed Point Formula). *For a separated scheme of finite type over $\mathbb{F}_q$, any $\overline{\mathbb{Q}}_\ell$-adic sheaf $\mathbb{L}$ on X, and every $m \neq 0$, we have*

$$\sum_{x \in X\left(\mathbb{F}_{q^{|m|}}\right)} \operatorname{tr}((\operatorname{Fr}_q \times 1)^m | \mathbb{L}_x) = \sum_i (-1)^i \operatorname{tr}((\operatorname{Fr}_q \times 1)^m | H_c^i(\overline{X}, \mathbb{L})).$$

Remark 6.4. (i) Since m is any nonzero integer, the formula holds also with the arithmetic Frobenius $1 \times \operatorname{Fr}_q$ instead of the geometric $\operatorname{Fr}_q \times 1$. (ii) Here X is not required to be smooth or proper. (iii) Underlying the proof is the observation that in characteristic $p > 0$ one has $\frac{d}{dx}(x^p) = 0$. Hence the graph of the Frobenius is transverse to the diagonal. In particular the fixed points of the Frobenius are isolated.

If $\overline{X}$ is proper and smooth over an algebraically closed field k, and $\mathbb{L}$ is a smooth $\overline{\mathbb{Q}}_\ell$-sheaf on $\overline{X}$, then a stronger variant of the fixed point formula (which is not used in this work) is known (see [SGA4 1/2], p. 151, for the case of the constant $\mathbb{L}$, and [SGA5], Exp. III, in general). To state this variant, let $i : \overline{X}' \hookrightarrow \overline{X} \times_k \overline{X}$ be a closed subscheme which is transverse to the diagonal morphism $\overline{\Delta} : \overline{X} \hookrightarrow \overline{X} \times_k \overline{X}$. Suppose that $f = \operatorname{pr}_1 \circ i$ is finite and flat. Put $h = \operatorname{pr}_2 \circ i$. Let $\mathbb{L} = (L_n)$ be a smooth $\overline{\mathbb{Q}}_\ell$-sheaf over $\overline{X}$ and $\beta : h^* \mathbb{L} \to f^* \mathbb{L}$ a sheaf morphism. Put $\alpha = a \circ \beta$. Then an endomorphism $H_c^i(f, \alpha, h)$ of $H_c^i(\overline{X}, \mathbb{L})$ is defined in (6.1.5) (for all i). For each point x' of $\overline{X}'$ we have $(h^* \mathbb{L})_{x'} = \mathbb{L}_{h(x')}$ by definition. If $h(x') = x$ and $f(x') = x$, then the sheaf morphism $\beta : h^* \mathbb{L} \to f^* \mathbb{L}$ induces a morphism $\beta_{x'} : (h^* \mathbb{L})_{x'} \to (f^* \mathbb{L})_{x'}$ on the stalks, namely $\beta_{x'}$ is an endomorphism of the finite dimensional $\overline{\mathbb{Q}}_\ell$-space $\mathbb{L}_x$. Then we have

Theorem 6.7 (Lefschetz Fixed Point Formula). *If $\overline{X}$ is proper and smooth over an algebraically closed field, and $\mathbb{L}$ is a smooth $\overline{\mathbb{Q}}_\ell$-sheaf on $\overline{X}$, then*

$$\sum_{\{x' \in \overline{X}'; h(x') = f(x') = x\}} \operatorname{tr}[\beta_{x'} | \mathbb{L}_x] = \sum_i (-1)^i \operatorname{tr}[H_c^i(f, \alpha, h) | H_c^i(\overline{X}, \mathbb{L})].$$

However, the scheme to which we are to apply the fixed point formula in Chap. 10 is not proper, and only the form of Theorem 6.6, where $f = \operatorname{id}$ and $\alpha = h = (\operatorname{Fr}_q \times 1)^m$ is a power of the Frobenius, is available. We shall now formulate a variant conjectured by Deligne, of the fixed point formula, and study its applications in Chaps. 11 and 12.

Theorem 6.8 (Deligne's Conjecture). *Suppose that X is a separated scheme of finite type over $\mathbb{F}_q$; $f, h : X' \to X$ morphisms, where h is proper and f is quasi-finite; $\mathbb{L}$ a smooth $\overline{\mathbb{Q}}_\ell$-adic sheaf on X; and $\alpha : h^* \mathbb{L} \to f^! \mathbb{L}$ a sheaf morphism which factorizes as the composition of a morphism $\beta : h^* \mathbb{L} \to f^* \mathbb{L}$ and* *the natural morphism* $a : f^*\mathbb{L} \to f^!\mathbb{L}$. *Then there exists an integer* m_0 *such that for any integer* m *with* $|m| \geq m_0$ *we have*

$$\sum_{x'} \mathrm{tr}[(\beta \circ (\mathrm{Fr}_q \times 1)^m)_{x'} | \mathbb{L}_x]$$

$$= \sum_i (-1)^i \, \mathrm{tr}[H_c^i(f, \alpha \circ (\mathrm{Fr}_q \times 1)^m, h \circ (\mathrm{Fr}_q \times 1)^m) | H_c^i(\overline{X}, \mathbb{L})].$$

On the left the sum ranges over all x' *in* $\overline{X}'$ *with* $(h \circ (\mathrm{Fr}_q \times 1)^m)(x') = f(x')$, *and we put* $x = f(x')$.

It suffices for us to assume that f is étale, in which case $f^* = f^!$.

This conjecture is motivated by the hope that after multiplication by a sufficiently high power of the Frobenius the correspondence $(f, h : X' \to X)$ becomes transverse to the diagonal $\Delta : X \hookrightarrow X \times_{\mathbb{F}_q} X$. The Lefschetz fixed point formula confirms the conjecture when X is proper and smooth, and Grothendieck's formula deals with the case of $f = h = \mathrm{id}$ and $\alpha = \mathrm{id}$.

Deligne-Lusztig (Ann. Math. 103 (1976), 103–161) noted that Deligne's conjecture holds for an automorphism of finite order of the scheme X. They multiplied the automorphism by a Frobenius and obtained a Frobenius (with respect to another structure on the scheme) for which the Grothendieck formula is again valid.

In fact Illusie [SGA5], Exp.III, Theorem 4.4, gives an explicit formula in terms of local data for the alternating sum on the cohomological (right) side of the formula, for any quasi-finite flat correspondence (loc. cit., (4.12), p. 111), and a complex $\mathbb{L}$ of sheaves in $D_c^b(\overline{X}, \overline{\mathbb{Q}}_\ell)$.

The problem is to compute these local terms. This was actually done in the case of a curve X, for a correspondence multiplied by a sufficiently high power of the Frobenius; the local data turned out to be the trace on the stalk $\overline{L}_x$, confirming Deligne's conjecture in the case of curves.

The usage of high powers of the Frobenius is already suggested by Drinfeld [D2], p. 166, ℓ. 1. Additional evidence is provided by the form of the trace formula which is proven in [FK2] and in Prop. 10.6. This form is a representation theoretic analogue of Deligne's algebro-geometric conjecture.

Remark 6.5. After the completion of the first draft of this work in 1983, several cases of Deligne's conjecture were proven by Pink [P] and Shpiz [Sp] (in a form not sufficiently strong as yet for our purposes: for a variety with smooth compactification by a divisor with normal crossings). They also reduced the conjecture to the conjectural resolution of singularities in positive characteristic.

Deligne's conjecture was finally proven unconditionally by Fujiwara [Fu] and by Varshavsky [V], by completely different techniques. We strongly recommend the lucid statement and proof of [V].

6.2. Congruence Relations.

6.2.1. We shall now return to our case of moduli schemes of elliptic modules. Thus let I be a nonzero ideal in A with $[V(I)] \geq 2$. Recall that $V(I)$ is the set of maximal ideals of A which contain I. By Theorem 4.9 the functor $\mathbb{F}_{r,I}$, which associates to any affine scheme Spec B over A the set of isomorphism classes of elliptic modules of rank r with structure of level I over B, is represented by an affine scheme $M_{r,I} = \operatorname{Spec} A_{r,I}$ of finite type over A. Let v be a maximal ideal of A. As usual we denote by U_v the maximal compact subgroup $\mathrm{GL}(r, A_v)$ of $G_v = \mathrm{GL}(r, F_v)$. Fix the Haar measure dg_v on G_v which assigns U_v the volume 1. For any open compact subgroup U'_v of U_v, let $\mathbb{H}(U'_v)$ denote the convolution $\overline{\mathbb{Q}}_\ell$-algebra of $\overline{\mathbb{Q}}_\ell$-valued compactly supported U'_v-biinvariant functions on G_v. Put $\mathbb{H}_v$ for $\mathbb{H}(U_v)$. As usual let U_I be the congruence subgroup of g in $G(\mathbb{A}_f)$ with $g - 1$ in $M(r, I\widehat{A})$. Then $U_{v,I} = G_v \cap U_I$ is equal to U_v for all v prime to I. The convolution algebra $\mathbb{H}_I$ of compactly supported $\overline{\mathbb{Q}}_\ell$-valued U_I-biinvariant functions on $G(\mathbb{A}_f)$ is isomorphic to the restricted direct product $\otimes_v \mathbb{H}(U_{v,I})$ of the local algebras $\mathbb{H}(U_{v,I})$; the isomorphism associates the characteristic function of $U_I g U_I$ with the product over v of the characteristic functions of $U_{v,I} g_v U_{v,I}$; note that $U_{v,I} g_v U_{v,I}$ is equal to U_v for almost all v.

An action of the adèle group $G(\mathbb{A}_f)$ on the moduli scheme $M_r = \varprojlim_I M_{r,I} = \varprojlim U_I \backslash M_r$ is defined in Prop. 4.14. Chapter 5 concerns the construction of a covering scheme $\widetilde{M}_r$ of M_r, with Galois group $D_\infty^\times$, such that $G(\mathbb{A}_f)$ acts on $\widetilde{M}_r$ and the action of $G(\mathbb{A}_f)$ on $\widetilde{M}_r$ commutes with the action of the Galois group. The group $G(\mathbb{A}_f)/F^\times$ acts on M_r without fixed points, and $M_{r,I}$ is smooth over $\mathbb{F}_p$. Put $\widetilde{M}_{r,I}$ for $U_I \backslash \widetilde{M}_r$, and let ρ be a finite dimensional representation of $\pi_1(\widetilde{M}_{r,I}/M_{r,I}) = D_\infty^\times$ with finite image. As noted in Example 6.3 in (6.1.4), to ρ one assigns a smooth $\overline{\mathbb{Q}}_\ell$-sheaf $\mathbb{L}(\rho)$ on $M_{r,I}$, and $\overline{\mathbb{Q}}_\ell$-adic cohomology spaces $H^i_{c,I} = H^i_c(\overline{M}_{r,I}, \mathbb{L}(\rho))$ and $H^i_{c,v,I} = H^i_c(\overline{M}_{r,I,v}, \mathbb{L}(\rho))$. Here $\overline{M}_{r,I}$ denotes the geometric generic fiber $M_{r,I} \times_A \overline{F}$ of $M_{r,I}$, and $\overline{M}_{r,I,v}$ is the geometric closed fiber $M_{r,I,v} \times_{\mathbb{F}_v} \overline{\mathbb{F}}_v$, where $M_{r,I,v} = M_{r,I} \times_A \mathbb{F}_v$ and $\overline{\mathbb{F}}_v$ is an algebraic closure of $\mathbb{F}_v = A/v$, for any $v \neq 0$ in Spec A. We also write $M_{r,v,I}$ for $M_{r,I,v}$.

The maps $\mathbb{H}_I \to \operatorname{End} H^i_{c,I}$ and $\mathbb{H}_I \to \operatorname{End} H^i_{c,v,I}$ of (6.1.8), which are induced by $g \mapsto H^i_{c,I}(g)$ and $g \mapsto H_{c,v,I}(g)$ (g in $G(\mathbb{A}_f)$), turn the $\overline{\mathbb{Q}}_\ell$-spaces $H^i_{c,I}$ and $H^i_{c,v,I}$ into $\mathbb{H}_I$-modules. In particular, if $v \neq 0$ in Spec A does not contain I, then $H^i_{c,I}$ and $H^i_{c,v,I}$ are $\mathbb{H}_v$-modules. Proposition 6.5 asserts that the $\mathbb{H}_I$-modules $H^i_{c,I}$ and $H^i_{c,v,I}$ are isomorphic for almost all $v \neq 0$ in Spec A, for all i.

6.2.2. Let v be a maximal ideal in A. An irreducible G_v-module π_v is called *unramified* if it contains a nonzero U_v-invariant vector, which is necessarily unique up to a scalar multiple. It is well known that there is a bijection between the sets of equivalence classes of (1) irreducible $\mathbb{H}_v$-modules in which the unit element acts as the identity and (2) irreducible unramified

G_v-modules: The irreducible G_v-module π_v defines a (one-dimensional) $\mathbb{H}_v$-module $\widetilde{\pi}_v$ by $\widetilde{\pi}_v(f_v) = \operatorname{tr} \pi_v(f_v)$ $(f_v \in \mathbb{H}_v)$, where $\pi_v(f_v)$ is the convolution operator $\int f_v(g)\pi_v(g)dg$, which factorizes through the projection on the one-dimensional subspace of U_v-fixed vectors in π_v. If I is a nonzero ideal in A then there is a bijection between the sets of equivalence classes of (1) irreducible $G(\mathbb{A}_f)$-modules π_f with a nonzero U_I-fixed vector and (2) irreducible $\mathbb{H}_I$-modules in which the unit element acts as the identity. It is given by $\pi_f \mapsto \pi_f^I$, where π_f^I is the space of U_I-fixed vectors in π_f.

Let S_r denote the symmetric group on r letters. It acts by permutation on $\overline{\mathbb{Q}}_\ell^{\times r}$. For each $\mathbf{z} = (z_i)$ $(1 \le i \le r)$ in $\overline{\mathbb{Q}}_\ell^{\times r}$ denote by $\chi_\mathbf{z}$ the unramified character $(b_{ij}) \mapsto \prod_i z_i^{\deg(b_{ii})}$ of the upper triangular subgroup B_v $(b_{ij} = 0$ if $i > j)$ of G_v. Let δ denote the character $\delta((b_{ij})k) = \prod_i |b_{ii}|_v^{r-2i+1}$ of G_v (here $k \in U_v = \operatorname{GL}(r, A_v)$). Let $I_v(\mathbf{z}) = \operatorname{Ind}(\delta^{1/2}\chi_\mathbf{z})$ denote the unramified G_v-module unitarily induced from $\chi_\mathbf{z}$. Its space consists of the locally constant functions $\mathbf{f} : G_v \to \overline{\mathbb{Q}}_\ell^\times$ with

$$\mathbf{f}(bg) = (\delta^{1/2}\chi_\mathbf{z})(b)\mathbf{f}(g) \qquad (b \in B_v, g \in G_v),$$

and G_v acts by right translation. The image of the representation $I_v(\mathbf{z})$ in the Grothendieck group, namely the equivalence class of its semisimplification, depends only on the projection of $\mathbf{z}$ in $\overline{\mathbb{Q}}_\ell^{\times r}/S_r$. This $I_v(\mathbf{z})$ has a unique irreducible unramified constituent $\widetilde{\pi}_v(\mathbf{z})$ in its composition series, and $(\widetilde{\pi}_v(\mathbf{z}))(f_v) = \operatorname{tr}(I_v(\mathbf{z}))(f_v)$ for all f_v in $\mathbb{H}_v$. Every irreducible $\overline{\mathbb{Q}}_\ell$-valued $\mathbb{H}_v$-module $\widetilde{\pi}_v$ is equivalent to $\widetilde{\pi}_v(\mathbf{z})$ for a unique $\mathbf{z}$ in $\overline{\mathbb{Q}}_\ell^{\times r}/S_r$. Moreover, the Satake homomorphism $f_v \mapsto f_v^\vee$, where $f_v^\vee(\mathbf{z}) = (\widetilde{\pi}(\mathbf{z}))(f_v)$, is an algebra isomorphism from $\mathbb{H}_v$ to $\mathbb{Q}[\overline{\mathbb{Q}}_\ell^{\times r}/S_r] = \mathbb{Q}[z_1, z_1^{-1}, \ldots, z_r, z_r^{-1}]^{S_r}$. In particular, $\mathbb{H}_v$ is commutative.

Remark 6.6. The commutativity of $\mathbb{H}_v$ quickly follows from the fact that the involution of $\mathbb{H}_v$ induced by the transpose map on G_v coincides with the identity on $\mathbb{H}_v$.

Let $\boldsymbol{\pi}$ denote a local uniformizer in the local ring A_v, and as usual put $q_v = |\boldsymbol{\pi}|_v^{-1}$. For any j $(1 \le j \le r)$ denote by g_j the diagonal matrix $(\boldsymbol{\pi}, \ldots, \boldsymbol{\pi}, 1, \ldots, 1)$ in G_v with $j = \deg(\det g_j)$. To simplify the notations choose the Haar measure on G_v which assigns U_v the volume one. Let ϕ_j be the characteristic function of the double coset $U_v g_j U_v$ in G_v. In (6.2.4) we use the following well-known lemma:

Lemma 6.9. *We have*

$$(\widetilde{\pi}_v(\mathbf{z}))(\phi_j) = q_v^{j(r-j)/2} \sum_{\underline{i}_j} z_{i_1} \ldots z_{i_j};$$

the sum ranges over all j-tuples $\underline{i}_j = (i_1, \ldots, i_j)$ of integers with $1 \le i_1 < i_2 < \cdots < i_j \le r$.

Proof. In the course of this proof we put $K = U_v$, $g = g_j$, $B = $ upper triangular subgroup of K. The double coset $B \backslash K / (K \cap gKg^{-1})$ is isomorphic to $B \backslash KgK/K$ by $k \mapsto kgK$. Moreover, $B \backslash K / (K \cap gKg^{-1})$ is isomorphic (on reducing modulo v) to $B(\mathbb{F}_v)\backslash G(\mathbb{F}_v)/P_j(\mathbb{F}_v)$, where $G = \mathrm{GL}(r)$ and P_j is the (lower triangular) parabolic subgroup of type $(j, r - j)$. We have $G(\mathbb{F}_v) = \cup_w B(\mathbb{F}_v)wP_j(\mathbb{F}_v)$, where the sum is disjoint and taken over $W/(W \cap P_j(\mathbb{F}_v))$. Here W is the Weyl group in $G(\mathbb{F}_v)$; $W \simeq S_r$, $W \cap P_j(\mathbb{F}_v) \simeq S_j \times S_{r-j}$, and the cardinality of $W/(W \cap P_j(\mathbb{F}_v))$ is $\binom{r}{j} = r!/j!(r-j)!$. This is the number of terms in the sum of the lemma. The double coset in $B\backslash KgK/K$ corresponding to $w = 1$ is BgK, and $\mathbf{f}(bgk) = \delta^{1/2}(g)z_1 \ldots z_j$ for any b in B and k in K. We have $\delta^{1/2}(g) = q_v^{j(r-j)/2}$. Since the symmetric group S_r permutes the monomials in $z_1, \ldots, z_r$ in the element $(\widetilde{\pi}_v(\mathbf{z}))(\phi_j)$ of $\mathbb{Q}[z_1, z_1^{-1}, \ldots, z_r, z_r^{-1}]^{S_r}$, each of the $\binom{r}{j}$ terms on the right of the formula of the lemma occurs in $(\widetilde{\pi}_v(\mathbf{z}))(\phi_j)$. Since there are only $\binom{r}{j}$ cosets in $W/(W \cap P_j(\mathbb{F}_v))$, there are only $\binom{r}{j}$ monomials in $(\widetilde{\pi}_v(\mathbf{z}))(\phi_j)$, and the lemma follows. $\qquad\square$

Our goal in this Sect. 6.2 is to prove the following. Let I be a nonzero ideal in A with $[V(I)] \geq 2$, $\widetilde{M}_{r,I}$ a finite étale Galois covering of $M_{r,I}$ contained in the Galois covering of $M_{r,I}$ with Galois group $D_\infty^\times$ which is constructed in Chap. 5, and ρ an irreducible $\overline{\mathbb{Q}}_\ell$-adic representation of the Galois group $\pi_1(\widetilde{M}_{r,I}/M_{r,I})$. Let $\widetilde{\pi}_f^I \otimes \sigma$ be an irreducible composition factor of the $\mathbb{H}_I \times \mathrm{Gal}(\overline{F}/F)$-module $H_c^i(\widetilde{M}_{r,I}, \mathbb{L}(\rho))$, $\widetilde{\pi}_v^I$ the component of $\widetilde{\pi}_f^I$ at v, and σ_v the restriction of σ to the decomposition subgroup $\mathrm{Gal}(\overline{F}_v/F_v)$. For all v with $H_{c,I}^i \simeq H_{c,v,I}^i$ (as $\mathbb{H}_I \times \mathrm{Gal}(\overline{\mathbb{F}}_v/\mathbb{F}_v)$-modules), σ_v factorizes via $\mathrm{Gal}(\overline{\mathbb{F}}_v/\mathbb{F}_v)$, and $\widetilde{\pi}_v^I$ is unramified. Let $\mathbf{z} = \mathbf{z}(\widetilde{\pi}_v)$ be an r-tuple $(z_j; 1 \leq j \leq r)$ in $\overline{\mathbb{Q}}_\ell^{\times r}$ whose image in $\overline{\mathbb{Q}}_\ell^{\times r}/S_r$ corresponds to the irreducible $\mathbb{H}_v$-module $\widetilde{\pi}_v^I$. Let $\mathrm{Fr}_v \times 1$ denote the geometric Frobenius morphism of $H_{c,v,I}^i$. We can now state the following theorem of congruence relations

Theorem 6.10. *For every i ($0 \leq i \leq 2(r - 1)$) and each irreducible composition factor $\widetilde{\pi}_f \otimes \sigma$ of $H_c^i(\overline{M}_{r,I}, \mathbb{L}(\rho))$ as an $\mathbb{H}_I \times \mathrm{Gal}(\overline{F}/F)$-module, and for every v with $H_{c,v,I}^i \simeq H_{c,I}^i$ as $\mathbb{H}_I \times \mathrm{Gal}(\overline{\mathbb{F}}_v/\mathbb{F}_v)$-modules, we have the following. Each eigenvalue u of the (geometric) Frobenius endomorphism $\sigma_v(\mathrm{Fr}_v \times 1)$ is equal to $q_v^{(r-1)/2}z_j$ for some $j = j(u)$ ($1 \leq j \leq r$).*

Corollary 6.11. *The (geometric) Frobenius endomorphism $\sigma_v(\mathrm{Fr}_v \times 1)$ has at most r distinct eigenvalues; they lie in the set $\{q_v^{(r-1)/2}z_j; 1 \leq j \leq r\}$.*

Remark 6.7.   (i) Using the trace formula, the purity (see [De3] or [SGA4 1/2]; Sommes Trig, p. 177/8) of the action of the Frobenius on $H_{c,v,I}^i$, and unitarity properties of the components of cuspidal automorphic representations of $G(\mathbb{A})$, we show in Chap. 10 that if the $D_\infty^\times$-module ρ corresponds to a cuspidal G_∞-module, then each z_j ($1 \leq j \leq r$), for every v which appears in Theorem 6.10, is algebraic with complex absolute values all equal to one. Hence the absolute value of each conjugate of the algebraic number u is $q_v^{(r-1)/2}$, independently of i.

(ii) It will be interesting to show that the dimension of the finite dimensional representation σ_v is bounded by, and moreover equal to, r.

(iii) The main geometric result of this section is the intrinsic "congruence relation" of correspondences of Prop. 6.13, which establishes an identity of correspondences on the scheme $M_{r,I,v}$. Its translation in (6.2.7) to a cohomological statement is formal. Consequently Theorem 6.10 is valid also when $H_{c,I}^i$ is replaced by cohomology H^i without compact support or any other cohomology theory.

6.2.3.   The proof of Theorem 6.10 will occupy the rest of this section. We begin with giving an alternative definition of the correspondence $T_j = T_{g_j}$ on the scheme $M_{r,v,I}$ associated as in (6.1.8) with the open compact subgroup U_v and the diagonal matrix $g_j = (\pi, \ldots, \pi, 1, \ldots, 1)$ in G_v, with $j = \deg(\det g_j)$.

Let $\mathbb{F}_j$ be the functor from the category of rings over A to the category of sets which associates to a ring B the set of isomorphism classes of elliptic modules φ of rank r over B, structures ψ of level I, and $\mathbb{F}_v$-module homomorphisms $\psi_j : \mathbb{F}_v^j \to E_v(B)$ with the property that the ideal $(\prod_u (x - \psi_j(u)); u \text{ in } \mathbb{F}_v^j)$ divides the ideal $J_v = (\varphi_a(x); a \text{ in } v)$ in $B[x]$. The proof of Theorem 4.10 shows that the functor $\mathbb{F}_j$ is representable by an affine scheme $M_{r,I,j} = \operatorname{Spec} A_{r,I,j}$. The ring $A_{r,I,j}$ is generated over $A_{r,I}$ by generators $c(u)$, u in $\mathbb{F}_v^j - \{0\}$, subject to the relations of the proof of Theorem 4.10. The group $G_j = \operatorname{GL}(j, \mathbb{F}_v)$ of automorphisms of $\mathbb{F}_v^j$ acts on $M_{r,I,j}$. Denote by $G_j \backslash M_{r,I,j}$ the quotient. It is equal to $\operatorname{Spec} A_{r,I,j}^{G_j}$.

Let h'_j be the morphism $M_{r,I,j} \to M_{r,I}$ defined by the finite flat generically separable (see Prop. 4.13) embedding $A_{r,I} \to A_{r,I,j}$; as a morphism of functors h'_j maps (φ, ψ, ψ_j) to (φ, ψ). Let f'_j be the morphism $M_{r,I,j} \to M_{r,I}$ defined as a morphism of functors by mapping (φ, ψ, ψ_j) to (φ', ψ); here φ' is the elliptic module defined by $P\varphi_a = \varphi'_a P$ for all a in A, where P is the endomorphism of the additive group with $P(x) = \prod_u (x - \psi_j(u))$ ($u \in \mathbb{F}_v^j$). Each of f'_j and h'_j is finite, flat, and étale outside v. Since they factorize through the quotient $M_{r,I,j} \to G_j \backslash M_{r,I,j}$, the correspondence S_j defined by $(f'_j, h'_j : M_{r,I,j} \to M_{r,I})$ (see (6.1.8)) is the multiple $|G_j| T_j$ of T_j.

6.2.4.   Put X for the fiber $M_{r,v,I} = M_{r,I,v} = M_{r,I} \otimes_A \mathbb{F}_v$ of $M_{r,I}$ at v, and similarly X_j for $M_{r,v,I,j} = M_{r,I,j} \otimes_A \mathbb{F}_v$. Denote by $f_j, h_j : X_j \to X$ the fibers at v of the morphisms f'_j and h'_j. It is clear that f_j, h_j are finite and flat, but not étale. Let (φ, ψ) denote the universal elliptic module of rank r with structure of level I, of characteristic v. Then $\varphi_a = \sum_i b_i(a) \tau^i$ ($b_i(a)$ in $A_{r,v,I}$) for all a in A. Fix a_0 in $v - v^2$. Then $b_0(a_0) = 0$ since φ has characteristic v. Define X^0 to be the open dense affine subscheme of X corresponding to the ring $A_{r,v,I}[x]/(x b_1(a_0) - 1)$. It is clear that the definition of X^0 is independent of the choice of a_0 in $v - v^2$. Denote by $\mathbb{F}^0$ the functor represented by X^0. The inverse images $X_j \times_X X^0$ of X^0 with respect to the two morphisms $f_j, h_j : X_j \to X$ coincide, and are equal to the subscheme X_j^0 of X_j defined by the requirement that the coefficient $b_1(a)$ of $\varphi_a = \sum_i b_i(a) \tau^i$ be invertible for all a in $v - v^2$. Denote by $\mathbb{F}_j^0$ the functor represented by X_j^0. The restrictions f_j^0, $h_j^0 : X_j^0 \to X^0$ of f_j, h_j are finite and flat by definition of X_j^0.

The ring A_j^0 such that $X_j^0 = \mathrm{Spec}\, A_j^0$ is generated over A^0 (where $X^0 = \mathrm{Spec}\, A^0$) by the elements $c(u)$ ($u \in \mathbb{F}_v^j - \{0\}$) subject to the relations of the proof of Theorem 4.10. In particular $\prod_u (x - c(u))$ ($u \in \mathbb{F}_v^j$) divides $\varphi_a(x)$ for all a in v. Since $c(0) = 0$, the product $\prod (x - c(u))$ over $u \neq 0$ in $\mathbb{F}_v^j$ divides $\varphi_a(x)/x$. Since $\varphi_a(x)/x = \sum_{i \geq 1} b_i(a) x^{q_v^i - 1}$, we conclude that for any a in $v - v^2$ and $u \neq 0$ in $\mathbb{F}_v^j$ the generator $c(u)$ satisfies the relation

$$\sum_{i \geq 1} b_i(a) c(u)^{q_v^i - 1} = c(u)^{q_v - 1} \sum_{i \geq 1} b_i(a) c(u)^{q_v^i - q_v} = 0.$$

Let X_j^{++} denote the closed subscheme of X_j^0 defined by the equation

$$\prod_{u \neq 0} c(u)^{q_v - 1} = 0;$$

the product ranges over all u in $\mathbb{F}_v^j - \{0\}$. The scheme X_j^{++} is a covering of degree $q_v - 1$ of the scheme X_j^+ which represents $\mathbb{F}_j^+$. The functor $\mathbb{F}_j^+$ is the functor which associates to each ring B over $\mathbb{F}_v$ the set $\mathbb{F}_j^+(B)$ consisting of all triples (φ, ψ, ψ_j) in $\mathbb{F}_j^0(B)$ with the property that $\psi_j : \mathbb{F}_v^j \to E_v(B)$ is not injective, that is, $\prod_{u \neq 0} c(u) = 0$. By the definition of X_j^0 the kernel of ψ_j is a line $\mathbb{F}_v$ in $\mathbb{F}_v^j$. Let X_j^- be the closed subscheme of X_j^0 defined by the equations

$$\sum_{i \geq 1} b_i(a) c(u)^{q_v^i - q_v} = 0 \qquad \text{for every } u \neq 0 \text{ in } \mathbb{F}_v^j.$$

It represents the subfunctor $\mathbb{F}_j^-$ of $\mathbb{F}_j^0$ such that $\mathbb{F}_j^-(B)$ consists of the (φ, ψ, ψ_j) in $\mathbb{F}_j^0(B)$ with injective $\psi_j : \mathbb{F}_v^j \to E_v(B)$. The intersection of X_j^{++} and X_j^- is empty. Indeed if $u \neq 0$ has $c(u)^{q_v - 1} = 0$ and $\sum_{i \geq 1} b_i(a) c(u)^{q_v^i - q_v} = 0$, then $b_1(a) = 0$, but $b_1(a)$ is a unit in X_j^0 (for a in $v - v^2$). We conclude that X_j^0 decomposes into two open closed disjoint components X_j^- and X_j^{++}, where the later is a covering of degree $q_v - 1$ of X_j^+. Consequently the maps $f_j^+, h_j^+ : X_j^+ \to X^0$ and $f_j^-, h_j^- : X_j^- \to X^0$, derived from $f_j^0, h_j^0 : X_j^0 \to X^0$, are finite and flat and étale. The morphisms f_j^0, h_j^0, f_j^{++}, and h_j^{++} are finite and flat, but not étale.

6.2.5. The triples $(f_j^+, h_j^+ : X_j^+ \to X^0)$, $(f_j^{++}, h_j^{++} : X_j^{++} \to X^0)$, and $(f_j^-, h_j^- : X_j^- \to X^0)$ define correspondences S_j^+, S_j^{++}, and S_j^- on the open dense subscheme X^0 of X. We have $S_j^{++} = (q_v - 1) S_j^+$. Since f_j^+, h_j^+ factorize through $G_j \backslash X_j^+$ and f_j^-, h_j^- through $G_j \backslash X_j^-$, we obtain correspondences T_j^+ and T_j^- on X^0 with $S_j^{++} = |G_j| T_j^+$ and $S_j^- = |G_j| T_j^-$. Since X_j^0 is the disjoint union of X_j^{++} and X_j^- we have $S_j = S_j^{++} + S_j^- = (q_v - 1) S_j^+ + S_j^-$, and $T_j = T_j^+ + T_j^-$.

Lemma 6.12. *We have an equality* $\mathrm{Fr}_v \circ T_j^- = q_v^j T_{j+1}^+$ *of correspondences on* X^0.

Proof. (i) Let $\widetilde{X}_j^-$ be the product of X_j^- and the set of surjective homomorphisms $\mathbb{F}_v^{j+1} \to \mathbb{F}_v^j$. It represents the functor $\widetilde{\mathbb{F}}_j^-$ which associates to a ring B over $\mathbb{F}_v$ the set of isomorphism classes of (1) elliptic modules φ of rank r over B with $b_1(a) \neq 0$ for the a in $v - v^2$, (2) structures ψ of level I, (3) injective $\mathbb{F}_v$-module homomorphisms $\psi_j : \mathbb{F}_v^j \to E_v(B)$, and (4) surjective homomorphisms $\gamma : \mathbb{F}_v^{j+1} \to \mathbb{F}_v^j$. The ideal $(\prod_u (x - \psi_j(u)); u \in \mathbb{F}_v^j)$ divides the ideal $J_v = (\varphi_a(x); a \in v)$ in $B[x]$. The morphism $\widetilde{h}_j^- : \widetilde{X}_j^- \to X^0$, $(\varphi, \psi, \psi_j, \gamma) \mapsto (\varphi, \psi)$, is finite and étale.

Define the morphism $\mathrm{Fr}_v \circ \widetilde{f}_j^- : \widetilde{X}_j^- \to X^0$ by $(\varphi, \psi, \psi_j, \gamma) \mapsto (\varphi', \psi)$, where the elliptic module φ' is defined by the relation $\tau^s P \varphi_a = \varphi'_a \tau^s P$ for all a in A. Here $s = \log_p q_v$, $P(x)$ is the polynomial $\prod_u (x - \psi_j(u))$ (u in $\mathbb{F}_v^j$), so that $\tau^s P$ is the polynomial $\prod_u (x - \psi_j(u))^{q_v}$. Then $\mathrm{Fr}_v \circ \widetilde{f}_j^-$ is finite and flat. Since the degree of $\widetilde{X}_j^-$ over X_j^- is $|G_j|(q_v^{j+1} - 1)/(q_v - 1)$, the correspondence determined by the triple $(\widetilde{h}_j^-, \mathrm{Fr}_v \circ \widetilde{f}_j^- : \widetilde{X}_j^- \to X^0)$ is

$$\frac{q_v^{j+1} - 1}{q_v - 1} |G_j| \, \mathrm{Fr}_v \circ S_j^- = \frac{q_v^{j+1} - 1}{q_v - 1} |G_j|^2 \, \mathrm{Fr}_v \circ T_j^-.$$

(ii) Let $\widetilde{\mathbb{F}}_{j+1}^+$ be the functor which associates to the ring B over $\mathbb{F}_v$ the set of isomorphism classes of (1) elliptic modules φ of rank r over B with $b_1(a) \neq 0$ for a in $v - v^2$, (2) structures ψ of level I, (3) $\mathbb{F}_v$-module homomorphisms $\psi_{j+1} : \mathbb{F}_v^{j+1} \to E_v(B)$ which are not injective, and (4) isomorphisms $\alpha : \mathbb{F}_v^j \overset{\sim}{\to} \psi_{j+1}(\mathbb{F}_v^{j+1})$. The partial level structure ψ_{j+1} is required to satisfy that $(\prod_u (x - \psi_{j+1}(u)); u$ in $\mathbb{F}_v^{j+1})$ divides $J_v = (\varphi_a(x); a \in v)$ in $B[x]$. As in the proof of Theorem 4.9 the functor $\widetilde{\mathbb{F}}_{j+1}^+$ is representable by an affine scheme $\widetilde{X}_{j+1}^+$. The morphism

$$\widetilde{X}_{j+1}^+ \to X_{j+1}^+, \quad (\varphi, \psi, \psi_{j+1}, \alpha) \mapsto (\varphi, \psi, \psi_{j+1})$$

is finite and étale, of degree $|G_j|$, since $G_j = \mathrm{Aut}\, \mathbb{F}_v^j$ acts on α.

The morphism

$$\widetilde{h}_{j+1}^+ : \widetilde{X}_{j+1}^+ \to X^0, \quad (\varphi, \psi, \psi_{j+1}, \alpha) \mapsto (\varphi, \psi),$$

is finite and flat. The morphism

$$\widetilde{f}_{j+1}^+ : \widetilde{X}_{j+1}^+ \to X^0, \quad (\varphi, \psi, \psi_{j+1}, \alpha) \mapsto (\varphi', \psi),$$

where φ' is the elliptic module determined by $P\varphi_a = \varphi'_a P$,

$$P(x) = \prod_u (x - \psi_{j+1}(u)) \quad (u \in \mathbb{F}_v^{j+1}),$$

is also finite and flat. It is clear that the correspondence determined by $(\widetilde{h}_{j+1}^+, \widetilde{f}_{j+1}^+ : \widetilde{X}_{j+1}^+ \to X^0)$ is

$$|G_j| S_{j+1}^+ = (q_v - 1)^{-1} |G_j| S_{j+1}^{++} = (q_v - 1)^{-1} |G_j| |G_{j+1}| T_{j+1}^+.$$

(iii) The map

$$\rho : (\varphi, \psi, \psi_j : \mathbb{F}_v^j \hookrightarrow E_v(B), \gamma : \mathbb{F}_v^{j+1} \twoheadrightarrow \mathbb{F}_v^j) \mapsto (\varphi, \psi, \psi_{j+1} = \psi_j \circ \gamma, \psi_j)$$

is a morphism from $\widetilde{\mathbb{F}}_j^-$ to $\widetilde{\mathbb{F}}_{j+1}^+$. To verify this claim, we have to show that $Q_{j+1}(x) = \prod_u (x - \psi_{j+1}(u))$ (u in $\mathbb{F}_v^{j+1}$) divides $\varphi_a(x)$ for all a in v. Our assumption is that $Q_j(x) = \prod_u (x - \psi_j(u))$ divides $\varphi_a(x)$ for all a in v. Since $\ker \gamma$ has cardinality q_v, we have that $Q_{j+1}(x) = Q_j(x)^{q_v}$. Now, since the separable additive polynomial $Q_j(x)$ divides $\varphi_a(x) = \sum_{i \geq 1} b_i(a) x^{q_v^i} = x^{q_v} R(x)$, it divides $R(x)$; consequently there is a polynomial P in $B[\tau]$ with $R = P Q_j$. Hence $\varphi_a = \tau^s R = \tau^s P Q_j = P' \tau^s Q_j = P' Q_{j+1}$, and $Q_{j+1}(x)$ divides $\varphi_a(x)$, as required. Here we put $s = \log_p q_v$ and $P' = \sum b_i^{q_v} \tau^i$ if $P = \sum b_i \tau^i$.

(iv) The map

$$(\varphi, \psi, \psi_{j+1} : \mathbb{F}_v^{j+1} \to B, \alpha : \mathbb{F}_v^j \xrightarrow{\sim} \psi_{j+1}(\mathbb{F}_v^{j+1})) \mapsto (\varphi, \psi, \alpha, \alpha^{-1} \circ \psi_{j+1})$$

defines a morphism from $\widetilde{\mathbb{F}}_{j+1}^+$ to $\widetilde{\mathbb{F}}_j^-$ which is inverse to the morphism ρ of (iv). It is clear that $\tilde{h}_{j+1}^+ \circ \rho = \tilde{h}_j^-$ and $\tilde{f}_{j+1}^+ \circ \rho = \mathrm{Fr}_v \circ \tilde{f}_j^-$. We conclude that the correspondences of the formulae displayed at the end of (i) and (ii) are equal. The lemma follows from the formula

$$|G_j| = \prod_{i=0}^{j-1} (q_v^j - q_v^i)$$

which implies that

$$|G_{j+1}| = (q_v^{j+1} - 1)|G_j| q_v^j.$$

$\square$

6.2.6. Congruence relation of correspondences.

Proposition 6.13. *We have the following equality of correspondences on* $M_{r,v,I}$:

$$\sum_{j \text{ odd}} q_v^{j(j-1)/2} \, \mathrm{Fr}_v^{r-j} \circ T_j = \sum_{j \text{ even}} q_v^{j(j-1)/2} \, \mathrm{Fr}_v^{r-j} \circ T_j \quad (0 \leq j \leq r).$$

Proof. Since the two morphisms $f_j, h_j : M_{r,v,I,j} \to M_{r,v,I}$ which define T_j are finite and flat, and X^0 is an open dense subscheme of $X = M_{r,v,I}$, it suffices to prove the displayed formula only for the restriction of the correspondences to X^0, namely to $f_j, h_j : M_{r,v,I,j}^0 \to M_{r,v,I}^0$. As a correspondence on X^0, each T_j decomposes as a sum of T_j^+ and T_j^-, so that the left side of the formula is the sum of I^+ and I^-, where

$$I^+ = \sum_{0 \leq j \leq r/2} q_v^{(2j+1)2j/2} \, \mathrm{Fr}_v^{r-2j-1} \circ T_{2j+1}^+$$

and

$$I^- = \sum_{0 \le j \le r/2} q_v^{(2j+1)2j/2} \, \mathrm{Fr}_v^{r-2j-1} \circ T_{2j+1}^-.$$

Note that $T_{r+1}^+ = 0 = T_{r+1}^-$, and $T_0^+ = 0$. Using Lemma 6.12 we rewrite I^- in the form

$$\sum_j q_v^{(2j+1)2j/2} q_v^{2j+1} \, \mathrm{Fr}_v^{r-2j-2} \circ T_{2j+2}^+ = \sum_j q_v^{(2j+1)(2j+2)/2} \, \mathrm{Fr}_v^{r-2j-2} \circ T_{2j+2}^+.$$

Hence the left side of the formula of the proposition is equal to

$$\sum_{0 \le j \le r} q_v^{j(j-1)/2} \, \mathrm{Fr}_v^{r-j} \circ T_j^+.$$

This is equal to the right side of the formula by the same argument, using Lemma 6.12, and our proposition follows. $\square$

6.2.7. It remains to prove Theorem 6.10.

Proof. By (6.1.9) the identity of Prop. 6.13 of correspondences on $M_{r,v,I}$ yields an identity of endomorphisms on the $\overline{\mathbb{Q}}_\ell$-module $H_c^i(\overline{M}_{r,I,v}, \mathrm{L}(\rho))$, as an $\mathbb{H}_v \times \mathrm{Gal}(\overline{\mathbb{F}}_v/\mathbb{F}_v)$-module, hence on each of its composition factors $\widetilde{\pi}_v^I \otimes \sigma_v$. On the factor $\widetilde{\pi}_v^I \otimes \sigma_v$ the correspondence T_j acts by $\widetilde{\pi}_v(\phi_j)$, where we denote by ϕ_j the characteristic function of the double coset $U_v g_j U_v$. By Lemma 6.9 the irreducible $\mathbb{H}_v$-module $\widetilde{\pi}_v = \widetilde{\pi}_v(z)$, $z = (z_j)$, satisfies

$$\widetilde{\pi}_v(\phi_j) = q_v^{(r-j)j/2} \sum_{\underline{i}_j} z_{i_1} z_{i_2} \ldots z_{i_j};$$

the sum ranges over all j-tuples $\underline{i}_j = (i_1, \ldots, i_j)$ with $1 \le i_1 < i_2 < \cdots < i_j \le r$. Applying the identity of Prop. 6.13 of correspondences to the factor $\widetilde{\pi}_v \otimes \sigma_v$ we obtain the identity

$$\sum_{j=0}^r (-1)^j q_v^{j(j-1)/2} \widetilde{\pi}_v(\phi_j) \sigma_v (\mathrm{Fr}_v \times 1)^{r-j} = 0.$$

This we rewrite in the form

$$\begin{aligned}
0 &= \sum_{j=0}^r (-1)^j q_v^{r(r-1)/2} \widetilde{\pi}_v(q_v^{j(j-r)/2} \phi_j) [q_v^{(1-r)/2} \sigma_v (\mathrm{Fr}_v \times 1)]^{r-j} \\
&= q_v^{r(r-1)/2} \sum_{j=0}^r (-1)^j \left(\sum_{\underline{i}_j} z_{i_1} \ldots z_{i_j} \right) [q_v^{(1-r)/2} \sigma_v (\mathrm{Fr}_v \times 1)]^{r-j}.
\end{aligned}$$

It remains to note that the characteristic polynomial p of a matrix Z whose eigenvalues are $z_1, \ldots, z_r$ is

$$p(t) = \det(tI - Z) = \sum_{j=0}^{r} (-1)^j \left(\sum_{i_j} z_{i_1} \ldots z_{i_j} \right) t^{r-j}.$$

Hence

$$p(q_v^{(1-r)/2} \sigma_v(\mathrm{Fr}_v \times 1)) = 0.$$

In particular, for each eigenvalue u of $\sigma_v(\mathrm{Fr}_v \times 1)$, we have

$$p(q_v^{(1-r)/2} u) = 0.$$

Hence $q_v^{(1-r)/2} u$ is equal to z_i for some i, and the theorem follows. $\qquad \square$

Part 3. Trace Formulae

The work of Part 4 depends on a comparison of the fixed point formula and the trace formula. Since only automorphic $G(\mathbb{A})$-modules occur in the Selberg formula, the purpose of this approach is to show that the $G(\mathbb{A}_f)$-modules $\tilde{\pi}_f$ which occur in the virtual module $H_c^* = \sum_i (-1)^i H_c^i$ are automorphic, in addition to establishing the relation concerning the local Frobenius and Hecke eigenvalues. The Grothendieck fixed point formula gives an expression for the trace of the action of the (geometric) Frobenius $\mathrm{Fr}_v \times 1$ on the cohomology module H_c^* by means of the set of points in $M_{r,I,v}(\overline{\mathbb{F}}_v)$ fixed by the action of the Frobenius and the traces of the resulting morphisms on the stalks of the $\overline{\mathbb{Q}}_\ell$-sheaf $\mathbb{L}(\rho)$ at the fixed points.

Part 3 prepares for the comparison. Following [D2], in Chap. 7 the set $M_{r,I,v}(\overline{\mathbb{F}}_v)$ is expressed as a disjoint union of isogeny classes of elliptic modules over $\overline{\mathbb{F}}_v$, and their types are studied. In Chap. 8 it is shown that the elliptic modules with level structure of a given type make a homogeneous space under the action of $G(\mathbb{A}_f)$, and the stabilizer is described. Moreover, the action of the Frobenius Fr_v is identified with multiplication by a certain matrix. A *type* is described in group theoretic terms of an elliptic torus in $G(F)$ (see Definition 7.3), and the cardinality of the set $M_{r,I,v}(\mathbb{F}_{v,n})$ ($[\mathbb{F}_{v,n} : \mathbb{F}_v] = n$) is expressed in terms of orbital integrals of conjugacy classes γ in $G(F)$ which are elliptic in $G(F_\infty)$ and n-admissible (see Definition 8.1) at v.

Next, in Chap. 9, it is shown that the orbital integral at v obtained in Chap. 8 can be expressed as an orbital integral of a spherical function $f_n = f_n^{(r)}$ on G_v whose normalized orbital integral $F(f_n)$ is supported on the n-admissible set. This spherical function is defined by the relation $\mathrm{tr}(\pi_v(z))(f_n) = q_v^{n(r-1)/2} \sum_{i=1}^r z_i^n$.

7. Isogeny Classes

The main tool which is applied in Part 4 is a comparison of the "arithmetic" fixed point formula with the "analytic" trace formula. To carry out this comparison we need to describe the arithmetic data, which is the cardinality of the set of points on the fiber $M_{r,v}$ at v of the moduli scheme M_r, over finite field extensions of $\mathbb{F}_v = A/v$, or, equivalently, the set $M_{r,v}(\overline{\mathbb{F}}_v)$ with the action of the Frobenius morphism on it, by group theoretic data which appears in the trace formula. In this Chapter we begin with a description (following Drinfeld [D2]) of the set of isogeny classes in $M_{r,v}(\overline{\mathbb{F}}_v)$ in terms of certain field extensions of F; these will be interpreted as tori of $\mathrm{GL}(r)$ in the trace formula.

Let d be a positive integer, and put $q = p^d$. Let B denote the ring $\mathbb{F}_q[\tau]$ generated by the indeterminate τ over $\mathbb{F}_q$ subject to the relation $u^p\tau = \tau u$ for all u in $\mathbb{F}_q$. The ring B is a domain, and its fraction ring $D = \mathbb{F}_q(\tau)$ is a division algebra of rank d (and dimension d^2) over its center $L = \mathbb{F}_p(t)$, where $t = \tau^d$. Then D is the cyclic division algebra sometimes denoted by $(\mathbb{F}_q(t)/\mathbb{F}_p(t), \tau, t)$, associated with the field extension $\mathbb{F}_q(t)/\mathbb{F}_p(t)$ and the element τ which acts as the Frobenius on $\mathbb{F}_q$, and satisfies $\tau^d = t$. Let R be the ring $\mathbb{F}_p[t]$ of functions in L regular in t.

Denote by v'' the place of L where $t = 0$ and by ∞ the place where $t^{-1} = 0$. At each place $w \neq v'', \infty$ of L we have that D is unramified, namely $D_w = D \otimes_L L_w$ is isomorphic over L_w to $M(d, L_w)$; moreover, $B_w = B \otimes_R R_w$ is isomorphic over L_w to $M(d, R_w)$; here R_w signifies the ring of integers of the completion L_w of L at w. At v'' and ∞ the division algebra D ramifies and has the invariants $1/d$ at v'', $-1/d$ at ∞. Indeed, in general, if E/F is a cyclic extension of local fields and σ generates $\mathrm{Gal}(E/F)$, the cyclic algebra $(E/F, \sigma, a) = \langle E, \sigma; \sigma x = \sigma(x)x, \sigma^{[E:F]} = a \rangle$, where $a \in F^\times$, has invariant $\mathrm{inv} = k/[E:F]$ in (the Brauer group $\mathrm{Br}(F) \simeq$) $\mathbb{Q}/\mathbb{Z}$ if a is π^k up to a unit in $F^\times$, where π is a uniformizer in $F^\times$.

Definition 7.1. An *ideal*, or *lattice*, in the division algebra D is a finitely generated R-module. An *ideal class* is the set of all right (or left) multiples by elements of $D^\times$ of an ideal. An *order* in D is a multiplicatively closed lattice, namely an open compact subring of D. If it is *maximal* then its tensor product with L over R is D. The *class number* of the division algebra D is the number of right (or left) ideal classes in any maximal order; it is independent of the choice of a maximal order. A *type* in D is an orbit under conjugation by $D^\times$ of a maximal order.

Proposition 7.1. *If B' is an order in D then there is x in $D^\times$ with*

$$xB'x^{-1} \subset B.$$

Proof. The ring B is a maximal order in D. Since $\mathbb{F}_q$ is perfect, D is Euclidean. Hence D has class number one. Consequently D has only one type, namely B is the unique maximal order in D up to conjugation by $D^\times$, as required. $\square$

Y.Z. Flicker, *Drinfeld Moduli Schemes and Automorphic Forms: The Theory of Elliptic Modules with Applications*, SpringerBriefs in Mathematics, DOI 10.1007/978-1-4614-5888-3_7, © Yuval Z. Flicker 2013

Corollary 7.2. *Let L' be a finite extension of L in D. Let R' be the ring of functions in L' which are regular outside ∞. Then there exists an x in $D^{\times}$ with $xR'x^{-1} \subset B$.*

Proof. The ring R' is an order in D.   □

Remark 7.1. A field extension L' of L embeds in D if and only if $[L' : L]$ divides d and both $L'_{v''} = L' \otimes_L L_{v''}$ and $L'_{\infty} = L' \otimes_L L_{\infty}$ are fields. The centralizer $Z_D(L')$ of L' in D is a division algebra, central of rank $d/[L' : L]$ over L'. The invariants of $Z_D(L')$ over L' are $[L' : L]/d$ at v'', $-[L' : L]/d$ at ∞, and 0 elsewhere.

Let $F = \mathbb{F}_q(C)$ be a function field as in Chap. 2, fix a place ∞ and denote by A the ring of functions on C which are regular on $C - \{\infty\}$. Let v be a fixed place in $\operatorname{Spec} A$, and fix $q = p^d$ as above.

Definition 7.2. (1) Let S denote the set of isogeny classes of elliptic modules φ of rank r and characteristic v over $\mathbb{F}_q$. (2) Let S' be the set of isomorphism classes of pairs (F', t), where F' is a field extension of F with $[F' : F]$ dividing r such that $F'_{\infty} = F' \otimes_F F_{\infty}$ is a field, and t is an element of $F'^{\times}$ with (i) $F' = F(t)$, (ii) $|t| = q^{1/r}$, and (iii) t has a zero only at one place v' of F' above v.

In (ii) the absolute value $|\cdot|$ is the extension to F'_{∞} of the absolute value on F_{∞} such that the valuation group $|F_{\infty}^{\times}|$ of $F_{\infty}^{\times}$ is $p^{\mathbb{Z}}$.

Theorem 7.3. *The map which associates to the elliptic module φ of rank r and characteristic v over $\mathbb{F}_q$ the pair (F', t), where $t\ (= \tau^d)$ is the Frobenius morphism in $\operatorname{End} E$ (we put E for E_{φ}), and F' is the subalgebra $F(t)$ of the algebra $F \otimes_A \operatorname{End} E$ generated by t, yields an isomorphism from S to S'.*

Proof. (i) Given a pair (F', t) in S', consider the maximal order $B = \mathbb{F}_q[\tau]$ in the division algebra $D = \mathbb{F}_q(\tau)$, where τ is an indeterminate subject to the relations $\tau^d = t$, and $u^p\tau = \tau u$ for all u in $\mathbb{F}_q$. We have $|t| = p^{d/r}$ by (ii). Let d'', r'' be relatively prime positive integers with $d/r = d''/r''$. The valuation group $|F^{\times}|$ of $F^{\times}$ is $p^{\mathbb{Z}}$. Since $\mathbb{Z} + d''\mathbb{Z}/r'' = \mathbb{Z}/r''$, the valuation group of $F'^{\times}$, which is generated by $p^{\mathbb{Z}}$ and $|t| = p^{d''/r''}$, is $p^{\mathbb{Z}/r''}$. The valuation group of $L = \mathbb{F}_p(t)$ is $p^{d''\mathbb{Z}/r''}$. Hence

$$
\begin{aligned}
[F' : F]/[F' : L] &= [F'_{\infty} : F_{\infty}]/[F'_{\infty} : L_{\infty}] \\
&= [\mathbb{Z}/r'' : \mathbb{Z}]/[\mathbb{Z}/r'' : d''\mathbb{Z}/r''] = r''/d'' = r/d,
\end{aligned}
$$

and $d/[F' : L] = r/[F' : F]$ is an integer. At the places v', ∞ of F' over the places v'', ∞ of L we have that $F'_{v'}$, F'_{∞} are fields. Hence F' embeds in the division algebra D which ramifies precisely at v'', ∞, and has rank d over L. In particular F embeds in D. Corollary 7.2 implies that after conjugating F by some x in $D^{\times}$, we may assume that the image of the ring A of functions in F regular outside ∞ lies in $B = \mathbb{F}_q[\tau]$.

To verify that the resulting homomorphism $\varphi : A \to B$ is an elliptic module, note that $|\varphi_a| = |a| = p^{v_p(a)} = |\tau|^{r v_p(a)}$, since $|\tau^d| = |t| = q^{1/r}$. Hence the highest power of τ in φ_a is $\tau^{r v_p(a)}$, and φ is an elliptic module of rank r whose characteristic is the restriction v of v' from F' to F.

(ii) Let $\varphi : A \to B = \mathbb{F}_q[\tau]$ be an elliptic module of rank r and characteristic v over $\mathbb{F}_q$. It extends to a homomorphism $F \to D = \mathbb{F}_q(\tau)$. The center of D is $L = \mathbb{F}_p(t)$, where $t = \tau^d$. Let $F' = F(t)$ be the subalgebra of $F \otimes_A \operatorname{End} E$ generated by the endomorphism t of E. Then F' is a commutative semisimple subalgebra of the division algebra D. Hence F' is a field. The division algebra D is ramified precisely at the two places v'', ∞ of L. The embedding of F' in D specifies two places v', ∞ of F' whose restrictions to L are v'', ∞, such that $F'_{v'}$ and F'_∞ are fields. The restrictions of v', ∞ to F are the places v, ∞. Indeed, the existence of the extension $A_v \hookrightarrow A'_{v'} \hookrightarrow D \otimes_R R_{v''} = \mathbb{F}_q[[\tau]]$ of $\varphi : A \to D = \mathbb{F}_q[\tau]$ implies that the characteristic of φ is v. Proposition 3.19 asserts that $F_\infty \otimes_A \operatorname{End} E$ is a division algebra. Hence $F' \otimes_F F_\infty$ is a field, and it is equal to F'_∞. On the other hand $F' \otimes_F F_v$ is the direct sum of $F'_{v'}$ and the $F'_{w'}$, where w' are the places of F' over v other than v'. Since the only zero of t in $L = \mathbb{F}_p(t)$ is by definition at v'', and v' is the only place of $F' = F(t)$ over v'', it follows that the only zero of t in F' is at v'.

Recall that $v_p(a), v_q(a)$ (a in A) are defined by $|a| = p^{v_p(a)} = q^{v_q(a)}$. As F lies in D we have $|a| = |\varphi_a| = |\tau^{r v_p(a)}| = |t^{r v_q(a)}|$. Hence $|\tau| = p^{1/r}$ and $|t| = q^{1/r}$. This defines the extension of the absolute value to F'. Since $|t| = q^{1/r}$ we have as in (i) that

$$r/[F' : F] = d/[F' : L] = \operatorname{rk}_L D/[F' : L].$$

This is an integer since F' is a field extension of L in D. $\qquad\square$

Remark 7.2. (i) The centralizer $D' = Z_D(F')$ of F' in D is a division algebra of rank $r' = r/[F' : F] (= d/[F' : L])$ over its center F'. Its invariants are $1/r'$ at v', $-1/r'$ at ∞, and 0 elsewhere. (ii) Since φ is defined over $\mathbb{F}_q$ we have that $\operatorname{End} E$ is an order in D. Since $F' = F(t)$ centralizes $\operatorname{End} E$, $\operatorname{End} E$ is an order in D'.

Let A' be the ring of functions in F' whose only possible pole is at ∞. As follows from (ii) in the proof of Theorem 7.3, and Corollary 7.2, the map $\varphi : A \to B = \mathbb{F}_q[\tau]$ extends to a monomorphism $\varphi' : A' \to B$, which is an elliptic A'-module of rank $r' = r/[F' : F]$ and characteristic v'. Since the class group of A' is finite, at each place $w' \neq \infty$ of F' we can choose π' in A' whose only zero is at w'. As in Theorem 3.4, we define the torsion module $E'_{\pi'^m}(\overline{\mathbb{F}}_p)$ where $E' = E_{\varphi'}$, and the Tate module

$$T_{w'}(E_{\varphi'}) = \operatorname{Hom}_{A'_{w'}}(F'_{w'}/A'_{w'}, \varinjlim_m E'_{\pi'^m}(\overline{\mathbb{F}}_p)).$$

At each $w' \neq v'$ in Spec A', the proof of Theorem 3.4 shows that $T_{w'}(E_{\varphi'})$ is a free $A'_{w'}$-module of rank r' (we simply have to replace A, F, v, r by A', F', v', r', etc.).

Proposition 7.4. *At v' we have $T_{v'}(E') = \{0\}$.*

Proof. We can use t to define $T_{v'}(E')$, since t lies in A' and the only zero of t is at v'. The torsion module $E'_{t^m}(\overline{\mathbb{F}}_p)$ consists of the x in $E'(\overline{\mathbb{F}}_q)$ ($= \overline{\mathbb{F}}_p$ as a group) with $t^m x = 0$ for some m. But t is a power of the Frobenius homomorphism $x \mapsto x^p$, and the only solution of $x^{p^m} = 0$ is $x = 0$. Hence $E'_{t^m}(\overline{\mathbb{F}}_p) = \{0\}$ for all m, and the proposition follows. □

Remark 7.3. Proposition 7.4 completes the description of the Tate module, which was started in Theorem 3.4 in the case where w is prime to the characteristic.

Let φ be an elliptic module of rank r and characteristic v over $\overline{\mathbb{F}}_p$. Since A is finitely generated and finite fields are perfect, φ is defined over $\mathbb{F}_q$ for some $q = p^d$ divisible by $q_v = |A/v|$. The construction of the pair (F', t) depends on the choice of d. If d is replaced by $d' \geq d$, then $t = \tau^d$ is replaced by $t' = \tau^{d'}$ and $F' = F(t)$ by $F'' = F(t')$. We shall associate to φ the smallest possible F' on considering all possible values of d, or t, namely on replacing t by a sufficiently large power of itself. We thus make the following:

Definition 7.3. An (F, v)-*type* is an isomorphism class of pairs (F', v') consisting of (i) a field extension F' of F such that $r' = r/[F' : F]$ is an integer and $F'_\infty = F' \otimes_F F_\infty$ is a field and (ii) a place v' of F' over v, such that for any t in A' with the property that t has a zero only at v', we have $F' = F(t)$.

Theorem 7.3 associates a unique type to each elliptic module φ over $\overline{\mathbb{F}}_p$.

Corollary 7.5. (*i*) *The map of Theorem 7.3 defines an isomorphism from the set of isogeny classes of elliptic modules of rank r and characteristic v over $\overline{\mathbb{F}}_p$ to the set of (F, v)-types (F', v'). (ii) The ring of endomorphisms of an elliptic module of type (F', v') is an order in a division algebra which is central over F' and whose nonzero invariants are $1/r'$ at v', $-1/r'$ at ∞. (iii) Moreover, $T_{w'}(E') = (A'_{w'})^{r'}$ for $w' \neq v'$ in Spec A', and $T_{v'}(E') = \{0\}$.*

Proof. Replacing φ by an isogenous elliptic module φ' amounts to conjugating F' by an element of $D^\times$. □

Proposition 7.6. *If (F', v') is an (F, v)-type, the completion $F'_{v'}$ of F' at v' is F_v.*

Proof. We have to show that the decomposition group of F'/F at v' is trivial, namely that any endomorphism σ of F' over F which fixes v' is trivial. Let t be a nonconstant element of A' whose only zeroes are at v'; its only pole is at ∞. Then $F' = F(t^m)$ for any positive integer m. Since $F'_\infty = F' \otimes_F F_\infty$ is a field we have $\sigma\infty = \infty$. Hence the only zeroes (resp. poles) of σt are at v' (resp. ∞), and they have the same multiplicity as those of t. Thus $t/\sigma t$ has neither zeroes nor poles. Hence it is a scalar in F'. Since the field of scalars of F' is finite, there exists a positive integer m with $\sigma t^m = t^m$. As F' is equal to $F(t^m)$, we conclude that σ fixes F', as required. $\qquad\square$

Let $Y = Y(F', v')$ denote the isogeny class of elliptic modules φ of rank r and characteristic v over $\overline{\mathbb{F}}_p$ of (F, v)-type (F', v'), equipped with a level structure $\psi : (F/A)^r \to E(\overline{\mathbb{F}}_p)$ of all levels. Then the set $M_{r,v}(\overline{\mathbb{F}}_p)$ of isomorphism classes of elliptic modules φ of rank r and characteristic v over $\overline{\mathbb{F}}_p$ with level structure ψ is the disjoint union of the Y over all types. Proposition 4.5 defines an action of the adèle group $G(\mathbb{A}_f) = \mathrm{GL}(r, \mathbb{A}_f)$ on $M_{r,v}(\overline{\mathbb{F}}_p)$.

Proposition 7.7. (i) *The group $G(\mathbb{A}_f)$ acts transitively on $Y = Y(F', v')$.*
(ii) *The Frobenius endomorphism $\mathrm{Fr}_v : x \mapsto x^{q_v}$ of $\mathbb{F}_v = A/v$ acts on Y.*

Proof. (i) Let $E = E_\varphi$, $E' = E_{\varphi'}$ be isogenous elliptic modules. Then there is an isogeny $P : E \to E'$. Denote its kernel by H. Via the level structure $\psi : (F/A)^r \to E(\overline{\mathbb{F}}_p)$ we identify H with the subgroup $\psi^{-1}H$ of $(F/A)^r$. Let g be an element of $G(\mathbb{A}_f) \cap M(r, \widehat{A})$ whose kernel is H when acting on $(F/A)^r$. Then gE is E'. We use the diagram

$$
\begin{array}{ccccccccc}
1 & \to & \psi^{-1}H & \to & (F/A)^r & \xrightarrow{\ g\ } & (F/A)^r & \to & 1 \\
 & & \downarrow \wr & & \psi \downarrow & & \downarrow g\psi & & \\
1 & \to & H & \to & E(\overline{\mathbb{F}}_p) & \xrightarrow{\ P\ } & E'(\overline{\mathbb{F}}_p) & \to & 1.
\end{array}
$$

Also, given E, and g in G, then gE is isogenous to E by definition.
(ii) This follows from Prop. 3.14(ii), with $p^j = q_v$ in the notations there. $\qquad\square$

Our aim in this remainder of this section is to fix notations and embeddings, relative to the fields F and F', which are used in the description of the isogeny class $Y(F', v')$ in Chap. 8. Denote by $\mathbb{A}'$ the ring of adèles of F', by $\mathbb{A}'_f$ the ring of adèles of F' without the ∞ component, and by $\mathbb{A}'^{v'}_f$ the F'-adèles without the ∞ and v' components. Similarly we have $\mathbb{A}$, $\mathbb{A}_f$, and $\mathbb{A}^v_f$ for F. Put $G_v = \mathrm{GL}(r, F_v)$ and $G^v = \mathrm{GL}(r, \mathbb{A}^v_f)$. Then $G(\mathbb{A}_f) = \mathrm{GL}(r, \mathbb{A}_f)$ is $G_v \times G^v$. Recall that $r' = r/[F' : F]$. To fix an embedding of $\mathrm{GL}(r', \mathbb{A}'_f)$ in $G(\mathbb{A}_f)$, note that the F-vector spaces $F'^{r'}$ and F^r are isomorphic. At each w in $\mathrm{Spec}\,A$ we then have an isomorphism over F_w of $\oplus_{w'} F'^{r'}_{w'} = (F' \otimes_F F_w)^{r'}$ with F^r_w; the sum ranges over all places w' of F' over w. We may choose the isomorphism so that $\oplus_{w'} A'^{r'}_{w'} = (A' \otimes_A A_w)^{r'}$ is mapped to A^r_w, namely we have $\oplus_{w'} T_{w'}(E') \simeq T_w(E)$. Further, we obtain an embedding of $\prod_{w'} \mathrm{GL}(r', F'_{w'})$ in $G_w = \mathrm{GL}(r, F_w)$, and we may regard the image as lying in a standard (diagonal) Levi subgroup. At $w = v$ we have $F' \otimes_F F_v = F'_{v'} \oplus (\oplus_{w'} F'_{w'})$, where w' ranges over the places of F' over v with $w' \neq v'$. Recall that $F'_{v'} \simeq F_v$ by Prop. 7.6. Consequently we have an embedding of $\mathrm{GL}(r', F'_{v'})$ as the group of matrices $\left(\begin{smallmatrix} g & 0 \\ 0 & 1 \end{smallmatrix}\right)$ in the standard Levi subgroup M_v of the standard (upper triangular) parabolic subgroup P_v of G_v of type $(r', r - r')$. The group P_v depends on F'. Denote by N_v the unipotent radical of P_v and by S_v the subgroup of M_v which is the image of the group of g in $\mathrm{GL}(r', F'_{v'})$ whose determinant $\det g$ is a unit in $F'_{v'}$.

Recall that $D'^{\times} = Z_D(F')^{\times}$ is the multiplicative group of a division algebra of rank r' central over F' which splits away from v', ∞. Fixing an isomorphism $D'_{w'} = D' \otimes_{F'} F'_{w'} \simeq M(r', F'_{w'})$ at each $w' \neq v', \infty$, we embed $D'^{\times}$ diagonally in $\mathrm{GL}(r', \mathbb{A}_f'^{v'})$, hence in $G^v = \mathrm{GL}(r, \mathbb{A}_f^v)$. At v' we have (i) that $D'_{v'} = D' \otimes_{F'} F'_{v'}$ is a division algebra and (ii) an epimorphism

$$D'_{v'}{}^{\times} \to \mathbb{Z} \simeq F'_{v'}{}^{\times}/A'_{v'}{}^{\times} \simeq \mathrm{GL}(r', F'_{v'})/S_v,$$

defined by the composition of the reduced norm and the valuation. Hence we have a map $D'^{\times} \to G_v/S_v$ which factorizes through M_v/S_v, obtained from combining all maps $D'_{w'} \to G_v/S_v$ for all places w' over v. In conclusion we obtain an embedding of $D'^{\times}$ in $G^v \times M_v/S_v$, hence in G/S_v.

8. Counting Points

We shall now describe each isogeny class in $M_{r,v}(\overline{\mathbb{F}}_p)$ and the action of the Frobenius on it. The group $G(\mathbb{A}_f)$ acts transitively on the isogeny class, and our task is to find the stabilizer of an element in the class, in order to describe the isogeny class as a homogeneous space. Recall that the isogeny class in $M_{r,v}(\overline{\mathbb{F}}_p)$ which corresponds to the type (F', v') is denoted by $Y(F', v')$. Denote by Fr_v the Frobenius morphism $x \mapsto x^{q_v}$ of $\overline{\mathbb{F}}_v$ over $\mathbb{F}_v = A/v$. As noted in Prop. 7.7 (ii), Fr_v acts on $Y(F', v')$. We use the notations M_v, N_v, S_v, etc., introduced at the end of Chap. 7. We shall now prove the following:

Proposition 8.1. (1) *Let* (φ, ψ) *be an elliptic module with level structure in the isogeny class* $Y(F', v')$ *in* $M_{r,v}(\overline{\mathbb{F}}_v)$. *Then the subgroup of* $G(\mathbb{A}_f)$ *which fixes the (isomorphism class of the) pair* (φ, ψ) *is equal to* $N_v S_v D'^{\times}$ *(the embedding of* $D'^{\times}$ *in* $G(\mathbb{A}_f)/S_v$ *is defined at the end of Chap. 7). Hence* $Y(F', v')$ *is isomorphic to* $(G(\mathbb{A}_f)/S_v N_v)/D'^{\times}$ *as a homogeneous space.* (2) *The Frobenius morphism* Fr_v *acts as right multiplication by an element* g *in* $G(\mathbb{A}'_f)$ $(\subset G(\mathbb{A}_f))$ *whose component* $g_{w'}$ *is 1 at each* $w' \neq v'$ *in* $\mathrm{Spec}\, A'$, *and* $g_{v'}$ *is an element of* $\mathrm{GL}(r', F'_{v'})$ *whose determinant is a uniformizer of* $F'_{v'} \simeq F_v$.

Proof. (i) Since $(F/A)^r = \oplus_w (F_w/A_w)^r$, the level structure $\psi : (F/A)^r \to E(\overline{\mathbb{F}}_p)$ is a set $\{\psi_w : (F_w/A_w)^r \to E_w(\overline{\mathbb{F}}_p)\}$ of level structures for all w in $\mathrm{Spec}\, A$. As usual, for each w we put $E_w = \varinjlim_m E_{\pi_w^m}$, where $E_{\pi_w^m}$ is the annihilator of π_w^m in E and π_w is an element of $A - \mathbb{F}_q$ whose only zero is at w. By Theorem 3.4, at each $w \neq v$, the level structure $\psi_w : (F_w/A_w)^r \to E_w(\overline{\mathbb{F}}_p)$ is an isomorphism; it defines an isomorphism of

$$\begin{aligned} A_w^r &= \mathrm{Hom}_{A_w}(F_w/A_w, (F_w/A_w)^r) \text{ with} \\ T_w(E) &= \mathrm{Hom}_{A_w}(F_w/A_w, E_w(\overline{\mathbb{F}}_p)). \end{aligned}$$

For g'_w in $G_w \cap M(r, A_w)$ and $a \neq 0$ in A_w, we clearly have that $g'_w \psi_w = a \psi_w$ if and only if $g'_w = a$, when $w \neq v$.

(ii) Recall from the end of Chap. 7 that at v we have

$$\begin{aligned} P_v &= M_v N_v \subset G_v, \quad M_v = M'_v M''_v, \quad M'_v = \mathrm{GL}(r', F_v), \\ M''_v &= \mathrm{GL}(r - r', F_v). \end{aligned}$$

Let g be an element of $G(\mathbb{A}'_f)$ with $g_{w'} = 1$ at $w' \neq v'$ and $g_{v'}$ in $M'_v \cap M(r, A_v)$. To analyze the action of g on (φ, ψ), recall that $T_{v'}(E') = \{0\}$; $E' = E_{\varphi'}$, and φ' is the elliptic A'-module extending φ from A to A'. The level structure map is therefore the zero map $\psi_{v'} : (F'_{v'}/A'_{v'})^r \to \{0\}$. For any nonscalar b in A' whose only zeroes are at v' we have that $\varphi'_b(x)$ is equal to a nonzero scalar multiple of $x^{|b|^r}$. Let n denote the valuation of the determinant of $g_{v'}$. If $n = 0$ then multiplication by $g_{v'}$ does not change the cardinality of $(A'_{v'}/bA'_{v'})^r$.

Y.Z. Flicker, *Drinfeld Moduli Schemes and Automorphic Forms: The Theory of Elliptic Modules with Applications*, SpringerBriefs in Mathematics, DOI 10.1007/978-1-4614-5888-3_8, © Yuval Z. Flicker 2013

Hence $g_{v'}\psi_{v'} = \psi_{v'}$. If $n = 1$, then the kernel of multiplication by $g_{v'}$ on $(F'_{v'}/A'_{v'})^{r'}$ is isomorphic to $\pi^{-1}A'_{v'}/A'_{v'}$; π denotes a uniformizer in $A'_{v'}$. Hence the element g acts on φ as the quotient map P on $\mathbb{G}_{a,\overline{\mathbb{F}}_p}$ whose kernel corresponds to the ideal (x^{q_v}) of $\overline{\mathbb{F}}_v[x]$. As noted in Prop. 7.7, this is the same as the action of Fr_v. For general n, g acts as Fr_v^n; (2) follows.

(iii) As noted at the end of Chap. 7, at v we have $T_v(E) \simeq A_v^{r-r'}$. The level structure ψ_v defines as in (i) a surjection $A_v^r \to A_v^{r-r'}$ with kernel $A_v^{r'}$. The level structure ψ_v is fixed by the group action (together with the elliptic module φ when this map is fixed). Note that any g'_v in $P_v = M'_v M''_v N_v$ can be written uniquely as a product $m'm''n$ with m' in M'_v, m'' in M''_v, and n in N_v. It follows that g'_v in $G_v \cap M(r, A_v)$ and $a \neq 0$ in A_v satisfy $g'_v\psi_v = a\psi_v$ if and only if g'_v lies in P_v, $m'' = a$, and $\deg(\det m') = r'\deg(a)$.

(iv) The action of $G(\mathbb{A}_f)$ on the isogeny class $Y(F', v')$ defines for each g' in $G(\mathbb{A}_f) \cap M(r, \widehat{A})$ an isogeny $\rho(g')$ in $B = \mathbb{F}_q[\tau]$ on the elliptic module $\varphi : A \to B$ (to another elliptic module φ' in $Y(F', v')$); for each g in $G(\mathbb{A}_f)$ we obtain an element $\rho(g)$ in the quotient ring $D = \mathbb{F}_q(\tau)$. It follows from (i) and (iii) that $\rho(g) = 1$ if and only if g lies in $S_v N_v$. Our purpose is to determine the set of g in $G(\mathbb{A}_f)/S_v N_v$ such that $\rho(g)$ is an element of $D = \mathbb{F}_q(\tau)$ which commutes with φ; in other words, φ' is φ. Namely, we have to determine the centralizer $Z_D(A)$ in D of (the image by φ of) A (in D). The center of D is L, and $F' = AL$. Hence $Z_D(A) = Z_D(F') = D'$. We conclude that $(G(\mathbb{A}_f)/S_v N_v)/D'^{\times}$ acts transitively on the set of points in $Y(F', v')$, as required.

$\square$

Let n be a positive integer. Let $\mathbb{F}_{v,n}$ denote the extension of degree n of $\mathbb{F}_v = A/v$ in $\overline{\mathbb{F}}_p$. The set $M_{r,v,I}(\mathbb{F}_{v,n})$ of $\mathbb{F}_{v,n}$-points of $M_{r,I,v}$, namely the set of isomorphism classes of elliptic modules of rank r, characteristic v, level I, over $\mathbb{F}_{v,n}$, is the set $M_{r,v,I}(\overline{\mathbb{F}}_p)^{\mathrm{Fr}_v^n}$ of the points in $M_{r,v,I}(\overline{\mathbb{F}}_p)$ fixed by Fr_v^n. It is the union over (F', v') of the sets $Y^{\mathfrak{F}}$, where $Y = Y(F', v')$ and $\mathfrak{F}$ in $\mathrm{GL}(r', F_v)$ has $\deg_v(\det \mathfrak{F}) = n$.

Let us describe the homogeneous space

$$Y = U\backslash(G(\mathbb{A}_f)/S_v N_v)/D'^{\times} = (U_v\backslash G_v/S_v N_v \times U^v\backslash G^v)/D'^{\times},$$

where $U = U_I$ is the congruence subgroup defined by I. By the Iwasawa decomposition

$$G_v = U_v P_v = U_v M_v N_v$$

we have

$$U_v\backslash G_v/N_v = U_v \cap M_v\backslash M_v,$$

and

$$U_v\backslash G_v/S_v N_v = \mathbb{Z} \times (U_v \cap M''_v\backslash M''_v),$$

as $M'_v/S_v = \mathbb{Z}$. Further,

$$D' \otimes_{F'} (F' \otimes_F F_v) = (D' \otimes_{F'} F'_{v'}) \oplus (\oplus_{w'} M(r', F'_{w'}))$$

(w' ranges over the primes of F' above v with $w' \neq v'$), and its multiplicative group maps onto $\mathbb{Z} \times M''_v$; we write $(v(\gamma), \gamma''_v)$ for the image in $\mathbb{Z} \times M''_v$ of γ in $D'^\times$; here $\gamma''_v = (\gamma, \ldots, \gamma)$ lies in the subgroup $\prod_{w'} \mathrm{GL}(r', F'_{w'})$ of M''_v. The group $D'^\times$ maps diagonally into $U^v \backslash G^v$. Hence

$$Y = [\mathbb{Z} \times (U_v \cap M''_v \backslash M''_v) \times (U^v \backslash G^v)]/D'^\times,$$

and $\mathfrak{F}$ is the element $(n, 1, 1)$ in Y. Note that $(n, 1, 1)$ commutes with (the image of) $D'^\times$.

Next we describe the points y in Y fixed by $\mathfrak{F}$. Thus $y\mathfrak{F} = y$. Then y is represented by a triple $g = (z, g_v, g^v)$ with z in $\mathbb{Z}$, g_v in M''_v, and g^v in G^v. Moreover, there are u_v in $U_v \cap M''_v$, u^v in U^v, and γ in $D'^\times$, such that $g\mathfrak{F} = (u_v u^v)^{-1} g\gamma$. Then $v(\gamma) = \deg_v(\det \gamma)$ equals n, where det denotes the reduced norm on the multiplicative group of the division algebra $D' \otimes_{F'} F'_{v'}$. Hence the equation $g\mathfrak{F} = (u_v u^v)^{-1} g\gamma$ becomes

$$z + n = z + v(\gamma), \quad g_v \gamma''_v g_v^{-1} = u_v \quad \text{and} \quad g^v \gamma (g^v)^{-1} = u^v.$$

To interpret these equations (in the next corollary), it will be convenient to introduce the following terminology. By a parabolic subgroup P of G_v we mean a standard one, one which contains the upper triangular subgroup. Given h in G_v there is a parabolic P such that the conjugacy class of h intersects (nontrivially) P, but not any proper parabolic subgroup of P. We call h *semisimple* if its conjugacy class intersects the (standard, i.e., containing the diagonal) Levi subgroup M of P. Note that this notion differs from the usual definition where h is required to be diagonalizable over an algebraic closure of F_v. For example, when $\mathrm{char}\, F_v = 2$ and b is in $F_v - F_v^2$, the element $h(b) = \left(\begin{smallmatrix} 0 & 1 \\ b & 0 \end{smallmatrix}\right)$ is semisimple in $\mathrm{GL}(2, F_v)$, but it is not diagonalizable over $\overline{F}_v$.

An equivalent definition is as follows: h is semisimple if the F_v-subalgebra H of the matrix algebra $M(r, F_v)$ generated by h is a semisimple algebra over F_v, namely H has no nonzero nilpotents. Then H is a direct sum of field extensions of F_v.

In general, let R be the nilradical of H. Then $\overline{H} = H/R$ is again a sum of fields. The problem of finding a Jordan decomposition for h is equivalent to the problem of lifting $\overline{H}$ to H, which is possible if $\overline{H}$ is separable over F_v but not in general (e.g., consider a nonperfect field F_v of characteristic 2, and take $a \in F_v - F_v^2$. Put $H = F_v[x]$ with $x^2 = a + y, y^2 = 0$. Then $\overline{H} = F_v[a^{1/2}]$, but there is no element in H with square a).

Every h in G_v has a Jordan decomposition as a product $h'h''$ of commuting semisimple and unipotent elements h' and h'' of G_v. An element h is called *elliptic* if the center of the centralizer $Z_h(G_v)$ of h in G_v is compact modulo the center Z_v of G_v. It is necessarily semisimple. If h is semisimple then $Z_h(G_v) \simeq \prod_i \mathrm{GL}(r_i, F_i)$, $1 \leq i \leq t$, where F_i are field extensions of F_v and $\sum_i r_i[F_i : F_v] = r$. Correspondingly, we write $h = h_1 \oplus \cdots \oplus h_t$. Each h_i

is elliptic in $\mathrm{GL}(r_i[F_i : F_v], F_v)$. In the example above, $h(b)$ is elliptic; its centralizer is the multiplicative group of the (inseparable) quadratic extension of F_v generated by $h(b)$. Let N_i be the norm map from F_i to F_v. Let n be an integer.

Definition 8.1. The semisimple element h of G_v is called n-*admissible* if there exists j $(1 \leq j \leq t)$ such that $\deg(N_j h_j) = n$ and $\deg(N_i h_i) = 0$ for all $i \neq j$. An element h of G_v is called n-admissible if its semisimple part is n-admissible.

There is a natural bijection from the set of conjugacy classes δ' in $(D' \otimes_{F'} F'_{v'})^\times$ to the set of elliptic conjugacy classes δ in $M'_v = \mathrm{GL}(r', F'_{v'})$. Here δ' corresponds to δ if they have equal characteristic polynomials. We conclude

Corollary 8.2. *The image in $(D' \otimes_{F'} F'_{v'})^\times$ of the element γ of $D'^\times$ defined above by y in $Y^{\mathfrak{F}}$ corresponds to an n-admissible element of G_v.* □

However, the element γ is not uniquely determined by y. We have the following:

Lemma 8.3. *The conjugacy class of γ in $D'^\times$ is uniquely determined by y.*

Proof. Suppose that the representative $g = (z, g_v, g^v)$ is replaced by $g' = u'g\delta$, u' in U, and δ in $D'^\times$. Then there are u'' in U, γ' in $D'^\times$ with $u'g\delta\mathfrak{F} = (u'')^{-1} \cdot u'g\delta \cdot \gamma'$. Use $g = u^{-1}g\gamma\mathfrak{F}^{-1}$ on the left to get $u'^{-1}u''u'u^{-1} = g(\delta\gamma'\mathfrak{F}^{-1}\delta^{-1}\mathfrak{F}\gamma^{-1})g^{-1}$. Hence $v(\gamma') = v(\gamma)$, and the element $\delta\gamma'\delta^{-1}\gamma^{-1}$ of $D'^\times$ has characteristic polynomial whose coefficients are integral at each w' in $\mathrm{Spec}\, A'$, and also at ∞, since the determinant of $\delta\gamma'\delta^{-1}\gamma^{-1}$ is rational in $F'^\times$ and a unit at each w' in $\mathrm{Spec}\, A'$. Hence $\delta\gamma'\delta^{-1}\gamma^{-1}$ is a scalar which is a unit in $F'^\times$. It has to be 1 since $U = U_I$ is a congruence subgroup and $I \neq \{0\}$ is prime to v. Hence $\gamma' = \delta^{-1}\gamma\delta$, as required. □

Proposition 8.4. *Let γ be an element of $D'^\times$ which is n-admissible over F_v. Let $Z_\gamma(D'^\times)$ be the centralizer of γ in $D'^\times$. The set of points y in $Y^{\mathfrak{F}}$ which correspond to γ is isomorphic to the set of cosets $g = (z, g''_v, g^v)$ in*

$$[\mathbb{Z} \times (U_v \cap M''_v \backslash M''_v) \times (U^v \backslash G^v)]/Z_\gamma(D'^\times)$$

with $g^v\gamma(g^v)^{-1}$ in U^v and $g''_v\gamma''_v(g''_v)^{-1}$ in $U_v \cap M''_v$.

Proof. We have to determine the set of g in $\mathbb{Z} \times (U_v \cap M''_v \backslash M''_v) \times (U^v \backslash G^v)$ which yield γ, namely satisfy (i) $u^{-1}g\gamma = g\mathfrak{F}$; here and below u, u', u'' lie in $(U_v \cap M''_v) \times U^v$. Replacing g by another representative $g' = u'g\delta$ $(\delta$ in $D'^\times)$ which also yields γ, thus $u''^{-1}g'\gamma = g'\mathfrak{F}$ or (ii) $u'^{-1}u''^{-1}u'g\delta\gamma = g\delta\mathfrak{F}$, we conclude from (i) and (ii) that $\delta\gamma = \gamma\delta$, and the proposition follows. □

Remark 8.1. If γ is an n-admissible element of G_v then it determines r', $M_v = M'_v \times M''_v$ and a field extension $F_v(\gamma)$ of F_v of degree r' and residual degree $e'_v(\gamma) = [\mathbb{Z} : \deg_v(\det Z_\gamma(M'_v))]$. Here det is the determinant in M'_v; $\det Z_\gamma(M'_v)$ is the image of the norm from $F_v(\gamma)^\times$ to $F_v^\times$. Further, $\deg_v$ is the valuation on $F'_{v'} = F_v$. The global element γ of $D'^\times$ determines the type (F', v'); see part (ii) of the proof of Prop. 8.5.

Definition 8.2. (i) The centralizer $G' = Z_\gamma(G)$ in G of a semisimple element γ in $G(F)$ is defined over F. An invariant differential form ω' of maximal degree on G' rational over F defines a Haar measure on $G'_w = Z_\gamma(G_w)$ at each place w of F and the product Haar measure on $G'(\mathbb{A}) = Z_\gamma(G(\mathbb{A}))$. Let Z_∞ be the center of G_∞. The volume

$$|Z_\gamma(G(\mathbb{A}))/Z_\gamma(G)Z_\infty| = |Z_\gamma(G_\infty)/Z_\infty| \cdot |Z_\gamma(G(\mathbb{A}_f))/Z_\gamma(G)|$$

is independent of the choice of the rational form. The differential form defines also a measure on any inner form of G'. Choose the measure so that the volume $|U_v \cap G'_v|$ is one. Similarly, for any semisimple γ in M''_v, we choose the Haar measure on $Z_\gamma(M''_v)$ with $|U_v \cap Z_\gamma(M''_v)| = 1$. In particular $|U_v \cap M''_v| = 1$. When $\gamma = 1$ we have $G' = G$ and the measure is denoted by ω. On discrete sets we choose the measure which assigns the value one to each point. Let dg be the product measure on $\mathbb{Z} \times M''_v \times G^v$.

(ii) Let χ^v be the quotient by $|U^v|$ of the characteristic function of U^v. For γ^v in G^v put

$$\Phi(\gamma^v, \chi^v) = \int_{G^v/Z_\gamma(G^v)} \chi^v(g\gamma g^{-1}) dg.$$

Let χ''_v be the characteristic function of $U_v \cap M''_v$ in M''_v. For γ''_v in M''_v put

$$\Phi(\gamma''_v, \chi''_v) = \int_{M''_v/Z_{\gamma''_v}(M''_v)} \chi''_v(g\gamma''_v g^{-1}) dg.$$

The measures are those of (i).

Proposition 8.5. *The set $M_{r,v,I}(\mathbb{F}_{v,n})$ is isomorphic to the union over all conjugacy classes of γ in $G(F)$ which are elliptic over F_∞ and n-admissible over F_v, of the cosets*

$$(z, g''_v, g^v) \in [\mathbb{Z} \times (U_v \cap M''_v \backslash M''_v) \times (U^v \backslash G^v)]/Z_\gamma(G)$$

with

$$\chi^v(g^v\gamma(g^v)^{-1})\chi''_v(g''_v\gamma''_v(g''_v)^{-1}) \neq 0.$$

The number of cosets corresponding to γ is

$$\int_{(\mathbb{Z} \times M''_v \times G^v)/Z_\gamma(G)} \chi''_v(g''_v\gamma''_v(g''_v)^{-1})\chi^v(g^v\gamma(g^v)^{-1}) dg$$
$$= |Z_\gamma(G(\mathbb{A}_f))/Z_\gamma(G)| \cdot e'_v(\gamma) \cdot \Phi(\gamma''_v, \chi''_v) \cdot \Phi(\gamma, \chi^v),$$

where $e'_v(\gamma) = [\mathbb{Z} : \deg_v(\det Z_\gamma(M'_v))]$ is defined in Remark 8.1.

Proof. (i) The set $M_{r,v,I}(\mathbb{F}_{v,n})$ is isomorphic to the union over all types (F', v') of the sets $Y^{\mathfrak{F}}$, $Y = Y(F', v')$, $\mathfrak{F} = \mathrm{Fr}^n_v$. Each set $Y^{\mathfrak{F}}$ is the disjoint union of sets parametrized by conjugacy classes γ in $D'^\times$, where D' is defined by F'. The conjugacy class γ in $D'^\times$ corresponds to a conjugacy class also denoted by γ in the centralizer $G' = Z_G(F')$ of $F'^\times$ in $G(F)$, hence in $G(F)$. This conjugacy class is elliptic in $G(F_\infty)$ and n-admissible in $G(F_v)$. In particular the centralizer $Z_\gamma((D' \otimes_{F'} F'_{v'})^\times)$ of γ in $(D' \otimes_{F'} F'_{v'})^\times$ is isomorphic to the centralizer $Z_\gamma(M'_v)$ of γ in M'_v, so that $e'_v(\gamma)$ is defined (see Remark 8.1). Moreover the inner forms $Z_\gamma(D'^\times)$ and $Z_\gamma(G')$ are isomorphic, and the conjugacy class γ intersects $(U_v \cap M''_v) \times U^v$.

(ii) Conversely, let γ be an element of $G(F)$ which is elliptic in $G(F_\infty)$, n-admissible in $G(F_v)$ (hence γ determines r' and a Levi subgroup $M_v = M'_v \times M''_v$ of G_v, which is standard, up to conjugacy), and its conjugacy class in $G(F)$ intersects $(U_v \cap M''_v) \times U^v$. Then the conjugacy class of γ determines a unique type (F', v'). Indeed, the definition of n-admissibility determines a place v'' of $F(\gamma)$ over v. The element γ is a unit outside v'', ∞, and its only zero is at v''. Let F' be the smallest field of the form $F(\gamma^m)$, where m is a positive integer. Let v' be the restriction of v'' to F'. Then (F', v') is a type which is uniquely determined by the conjugacy class of γ, and the proposition follows.

$\square$

Our next aim is to express the factor $e'_v(\gamma)\Phi(\gamma''_v, \chi''_v)$ and the condition that γ is n-admissible, in a unified, convenient way. We define in the next section a spherical function $f_{n,v}$ on G_v whose orbital integral $\Phi(\gamma, f_{n,v})$ will turn out to have the property that it is zero unless γ is n-admissible in G_v, where it is equal to our factor.

9. SPHERICAL FUNCTIONS

In this chapter we compute the orbital integrals of a certain spherical function, which is introduced in Definition 9.1. We give two methods of computation. That of Prop. 9.9 is natural; it is based on representation theoretic techniques, as presented, e.g., in [BD, BZ, Bo, C, F2, FK1, K1, K2]. That of Prop. 9.12 is elementary. It is due to Drinfeld. This chapter is independent of the rest of the book. In particular, we book with a local field F which is non-Archimedean but of any characteristic.

We begin with fixing the notations. Let R be the ring of integers of F; $G = \mathrm{GL}(r, F)$; $K = \mathrm{GL}(r, R)$; Z the center of G; A the diagonal subgroup; U the upper triangular unipotent subgroup; $B = AU$; and $P = MN$ a maximal parabolic subgroup of G of type $(s, r - s)$, with Levi subgroup $M = M' \times M''$ ($M' = \mathrm{GL}(s, F), M'' = \mathrm{GL}(r - s, F)$) containing A, and unipotent radical N contained in U. Write P_s, M'_s, etc., to indicate the dependence on s ($1 \le s \le r$).

Let π be a local uniformizer. Put $q = |R/\pi R|$. Normalize the absolute value $|\cdot|$ and the valuation deg by $|\pi| = q^{-1}$ and $|a| = q^{-\deg(a)}$. All Haar measures are normalized here to assign the volume one to the intersection with K.

Let $C_c(G//K)$ (resp. $C_c(Z\backslash G//K)$) be the convolution algebra of *spherical*, namely complex-valued K-biinvariant compactly supported functions f, on G (resp. G/Z). Similarly we have $C_c(M//M \cap K)$. The map $f \mapsto \overline{f}$, $\overline{f}(x) = \int_Z f(zx)dz$, takes $C_c(G//K)$ onto $C_c(Z\backslash G//K)$.

Let x be a regular element of G. If the eigenvalues of x are $x_1, \ldots, x_r$, put

$$\Delta(x) = \left| \prod_{i<j} (x_i - x_j)^2 / x_i x_j \right|^{1/2}.$$

For $x = (x', x'')$ in $M = M' \times M''$ put $\Delta_M(x) = \Delta_{M'}(x')\Delta_{M''}(x'')$.

Let T be the centralizer of x in G and $'T$ its split component. Put

$$\Phi(x, f) = \int_{G/T} f(gxg^{-1})dg, \qquad F(x, f) = \Delta(x)\Phi(x, f),$$

$$'\Phi(x, f) = \int_{G/'T} f(gxg^{-1})dg, \qquad 'F(x, f) = \Delta(x)'\Phi(x, f).$$

For x in M, f on M, put $\Phi^M(x, f) = \int_{M/T} f(gxg^{-1})dg$, etc.

For f on G put $f_N(x) = \delta_P(x)^{1/2} \int_N \int_K f(k^{-1}xnk)dndk$. Then $F(x, f) = F^M(x, f_N)$ for x in M (see, e.g., [FK1], Chap. 7). For $x = (x_1, \ldots, x_r)$ in A with $m_i = \deg(x_i)$ we put $m = (m_1, \ldots, m_r)$ and $F(m, f) = q^{1/2\langle \sum \alpha, m\rangle} \int_U f(xu)du$; the sum runs through all positive roots α of A in U. For f in $C_c(G//K)$ we then have that $F(x, f) = F(m, f)$; namely $F(x, f)$ depends only on the valuations of the eigenvalues of x.

Y.Z. Flicker, *Drinfeld Moduli Schemes and Automorphic Forms: The Theory of Elliptic Modules with Applications*, SpringerBriefs in Mathematics, DOI 10.1007/978-1-4614-5888-3_9, © Yuval Z. Flicker 2013

For f in $C_c(G//K)$ we define the Satake transform $f^\vee$ of f to be the polynomial $f^\vee(z) = \sum_m F(m, f)z^m$ in $z_1, \ldots, z_r, z_1^{-1}, \ldots, z_r^{-1}$; m runs through $\mathbb{Z}^r$; $z = (z_1, \ldots, z_r)$ varies over $\mathbb{C}^{\times r}$; and we put $z^m = z_1^{m_1} \ldots z_r^{m_r}$. The symmetric group S_r on r letters acts on $\mathbb{C}^{\times r}$ by permuting the indices of the entries of z. The theory of the Satake transform asserts that the map $f \mapsto f^\vee$ is an isomorphism from the algebra $C_c(G//K)$ to $\mathbb{C}(\mathbb{C}^{\times r}/S_r)$. This isomorphism has the following alternative description. We first note that the set of unramified irreducible G-modules is isomorphic to $\mathbb{C}^{\times r}/S_r$ as follows. To z corresponds a unique irreducible unramified constituent $\pi(z)$ in the G-module $I(z)$ normalizedly induced from the character $(b_{ij}) \mapsto \prod_i z_i^{\deg(b_{ii})}$ of B. The elements $z_1, \ldots, z_r$ are called the *Hecke eigenvalues* of $\pi(z)$, or $I(z)$. Then $C_c(G//K)$ is isomorphic to the algebra $\mathbb{C}(\mathbb{C}^{\times r}/S_r)$ by $f^\vee(z) = \mathrm{tr}(\pi(z))(f)$. In particular, to define a spherical function, it suffices to define its Satake transform, or equivalently its orbital integrals on the split set of G.

Definition 9.1. Let $f_n = f_n^{(r)}$ be the member of $C_c(G//K)$ defined by $\mathrm{tr}(\pi(z))(f_n) = q^{n(r-1)/2} \sum_{i=1}^r z_i^n$; equivalently, f_n is defined by $F(m, f_n) = q^{n(r-1)/2}$ if $m = (n, 0, \ldots, 0)$ in $\mathbb{Z}^r/S_r$, and $F(m, f_n) = 0$ otherwise.

For x in $G = \mathrm{GL}(r, F)$, put $v(x) = \deg(\det x)$; it is an integer, in $\mathbb{Z}$. For x in $\mathrm{PGL}(r, F)$, define $v(x)$ in $\mathbb{Z}/r\mathbb{Z}$ to be $v(x') \bmod r$, where x' is a representative of x in G. The superscript (r) in $f^{(r)}$ is to emphasize that this is a function on $\mathrm{GL}(r, F)$.

Proposition 9.1. *The normalized orbital integral $F(f_n)$ is supported on the n-admissible (see Definition 8.1) set of G.*

Proof.   (i) Let T_n be the set of x in G with $v(x) = \deg(\det x)$ equals n. Then the integral $F(f_n)$ is zero outside T_n. Indeed, the set T_n is K-biinvariant, and the orbital integrals of f_n multiplied by the characteristic function of T_n are equal to those of f_n on A. Consequently f_n itself is supported on T_n.

(ii) Given s we have $P = P_s = MN$ and a characteristic function χ of the set of $x = (x', x'')$ in M with $|\det x'| \leq |\det x''|$ (thus $v(x') \geq v(x'')$). The function $\chi f_{nN}^{(r)}$ lies in $C_c(M//M \cap K)$. We claim that it is equal to the function $q^{n(r-s)/2} f_n^{(s)}(x') f_0^{(r-s)}(x'')$ in $C_c(M//M \cap K)$. Indeed, for any $m = (m', m'')$ in $\mathbb{Z}^r$, where $m' = (m'_i)$ in $\mathbb{Z}^s$, $m'' = (m''_j)$ in $\mathbb{Z}^{r-s}$, we have

$$F^M(m, \chi f_{nN}^{(r)}) = F^M(m, f_{nN}^{(r)}) = F(m, f_n^{(r)})$$

$$= q^{n(r-s)/2} F^{M'}(m', f_n^{(s)}) F^{M''}(m'', f_0^{(r-s)})$$

if $\sum_i m'_i \leq \sum_j m''_j$, by definition, and

$$F^M(m, \chi f_{nN}^{(r)}) = 0 = F^{M'}(m', f_n^{(s)}) F^{M''}(m'', f_0^{(r-s)})$$

otherwise. This proves the claim. Since

$$F^M((x', x''), \chi f_{nN}^{(r)}) = F((x', x''), f_n^{(r)})$$

on the $x = (x', x'')$ with $\chi(x) \neq 0$, the proposition follows by induction on $s \geq 1$.

$\square$

Definition 9.2. A (locally constant compactly supported complex-valued) function f on G is called *discrete* if its orbital integral $\Phi(x, f)$ vanishes at each regular (distinct eigenvalues) nonelliptic element x of G.

Corollary 9.2. *For every $n \geq 1$ there exist a discrete function f_n^{disc} on G and a function f_n^{reg} on G which vanishes on the elliptic set of G, such that $\Phi(x, f_n) = \Phi(x, f_n^{\mathrm{disc}}) + \Phi(x, f_n^{\mathrm{reg}})$ for every x in G.*

Proof. Note that all eigenvalues of an n-admissible element have the same valuations if and only if the element is elliptic and if and only if none of its eigenvalues is a unit. Since the valuation group is discrete and the topology on G is totally disconnected, it is clear that there exists an open closed subset T of G which contains all n-admissible x in G which are not elliptic. The corollary holds with $f_n^{\mathrm{reg}} = \theta f_n$, where θ is the characteristic function of T. $\square$

By virtue of Prop. 9.1, to compute $\Phi(x, f_n)$ it suffices to consider n-admissible x. In this case we may assume that $x = (x', x'')$ lies in $M = M' \times M''$, where x' is elliptic in M' with $v(x') = n$, and x'' lies (up to conjugation in M'') in $M'' \cap K = \mathrm{GL}(r - s, R)$. Note that the eigenvalues of x'' are units, but those of x' are not. In this case

$$\Delta_M(x)/\Delta(x) = D_{G/M}(x)^{-1/2} = \delta_P(x)^{-1/2} = q^{-n(r-s)/2}$$

(see, e.g., [FK1], Prop. 7). Hence

$$\begin{aligned}
\Phi(x, f_n^{(r)}) &= (\Delta_M(x)/\Delta(x))\Phi^M(x, f_{nN}^{(r)}) = q^{-n(r-s)/2}\Phi^M(x, f_{nN}^{(r)}) \\
&= q^{-n(r-s)/2}[q^{n(r-s)/2}\Phi^{M'}(x', f_n^{(s)})\Phi^{M''}(x'', f_0^{(r-s)})] \\
&= \Phi^{M'}(x', f_n^{(s)})\Phi^{M''}(x'', f_0^{(r-s)}).
\end{aligned}$$

Since the second factor here is $\Phi(x'', \chi_v'')$ in the notations of Prop. 8.5, this reduces the computation to the case where x is elliptic; here $r = s$ and $x = x'$.

Let G be a reductive p-adic group with compact center. A unitary (by which we mean unitarizable) irreducible admissible G-module π_0 is called *square-integrable* if its matrix coefficients are absolutely square integrable on G. Let π_0 be a square integrable G-module.

Definition 9.3. A compactly supported locally constant complex-valued function f_0 on G is called a *pseudo-coefficient* of π_0 if $\mathrm{tr}\, \pi_0(f_0) = 1$ and $\mathrm{tr}\, \pi(f_0) = 0$ for every irreducible tempered G-module π inequivalent to π_0.

Although the following is valid in greater generality, we state it only in our context of $G = \mathrm{GL}(r, F)$.

Lemma 9.3.   (1) *For any square-integrable π_0 there exists a pseudo-coefficient.*

(2) *The orbital integral $\Phi(x, f_0)$ of a pseudo-coefficient f_0 is zero on every regular nonelliptic element x of G, and $'\Phi(x, f_0) = \overline{\chi}_{\pi_0}(x)$ on the regular elliptic set of G; here $\overline{\chi}_{\pi_0}$ is the complex conjugate of the character χ_{π_0} of π_0.*

Proof.   (1) Follows from the trace Paley–Wiener theorem of [BDK].

(2) This is Theorem K of [K1] in characteristic zero. When char F is positive, denote by G' an anisotropic form of G. Note that one has a correspondence of G-modules with G'-modules, stated by means of character relations; this well-known correspondence can be deduced using [K2] from the analogous result in characteristic zero. Alternatively, a simple proof of this correspondence, also in positive characteristic, is given in [F6]. From the corresponding standard fact for G'-modules, it then follows that the restrictions of the characters of the square-integrable G-modules to the elliptic set make a complete orthonormal set in the usual inner product on the set of class functions on the elliptic regular set of G. Since the character is a locally constant function on the regular set (a well-known result of Harish-Chandra), the orbital integral of a pseudo-coefficient f_0 of π_0 is related to χ_{π_0} as in (2).   □

For each $s \leq r$, let St_s be the Steinberg $\mathrm{PGL}(s, F)$-module contained in $I_s = I(q^{(s-1)/2}, q^{(s-3)/2}, \ldots, q^{(1-s)/2})$. By [Z] it is the unique square-integrable constituent in I_s. Let ω_s be an unramified character of $F^\times$ primitive of order s, and put $\zeta_s = \omega_s(\boldsymbol{\pi})^{-1}$. Let $f_{s,i}$ be a pseudo-coefficient of $\mathrm{St}_s \otimes \omega_s^i$; it exists by Lemma 9.3. Put $u_{n,s}(x) = \sum_{i=1}^s \zeta_s^{ni} f_{s,i}(x)$.

Lemma 9.4. *If x is regular then $'\Phi(x, u_{n,s})$ is zero unless x is elliptic with $v(x) \equiv n(\mathrm{mod}\, s)$, where it is equal to $(-1)^{s-1}s$.*

Proof. This follows from Lemma 9.3(2), since the character of St_s is $(-1)^{s-1}$ on the elliptic set.   □

Let J be an Iwahori subgroup of G. Then we have the following.

Lemma 9.5. *An irreducible G-module π has a nonzero J-fixed vector if and only if it is a subquotient of an induced unramified G-module $I(z)$.*

Proof. This is (4.7) in [Bo]. Another proof, and a generalization to the tame subgroup, is in [F7], Theorem 2.1.   □

Corollary 9.6. *The representation $\mathrm{St}_{r,i}$ has a nonzero J-fixed vector.*

Proposition 9.7. *Let π_0 be a square-integrable G-module which has a nonzero J-fixed vector. Then π_0 has a pseudo-coefficient f_0 with the property that $\mathrm{tr}\,\pi(f_0) = 0$ for every irreducible π which has no nonzero J-fixed vector.*

Proof. If f_{π_0} is a pseudo-coefficient of π_0 and 1_J is the characteristic function of J, then $f_0 = 1_J * f_{\pi_0}$.   □

Remark 9.1. By the results of [Bo, BD], the compact open subgroup J is "good" in the following sense. The category C of algebraic G-modules decomposes as the direct sum of the category C_J of G-modules whose subquotients all have nonzero J-fixed vectors and the category C^J of G-modules whose subquotients never have nonzero J-fixed vectors. Let $Z(C) = Z(C_J) \oplus Z(C^J)$ be the corresponding decomposition of the Bernstein co-center (see [B, BD]) of this category. Let $H(G)$ denote the convolution algebra of compactly supported locally constant functions on G. By [BD] there exists f_1 in $H(G)$ which acts trivially on C_J and as zero on C^J. Let f_0' be any pseudo-coefficient of π_0. Then $f_0 = f_0' * f_1$ has the properties asserted by the proposition.

Corollary 9.8. *There exists a pseudo-coefficient $f_{r,i}$ of $\mathrm{St}_r \otimes \omega_r^i$ with the following property. Every irreducible π with $\mathrm{tr}\,\pi(f_{r,i}) \neq 0$ has a nonzero J-fixed vector.*

The main result of this section is the following:

Proposition 9.9. *Suppose that x is elliptic. Then $'\Phi(x, f_n^{(r)})$ is r if $v(x) = n$, and 0 otherwise.*

Proof. To prove the proposition it suffices to show that for x elliptic, $'\Phi(x, \overline{f}_n^{(r)})$ is zero unless $v(x) \equiv n(\mathrm{mod}\, r)$, where the value is r. This will now be proven in five steps, when x is elliptic regular. Then we will reduce the case of any elliptic x to this special case.

   (i) By induction we assume the claim for all $s < r$, and prove it for r. For each $s < r$, let $\overline{h}_{n,s}$ be a compactly supported locally constant function on $Z \backslash G$ such that for all regular x we have $'\Phi(x, \overline{h}_{n,s}) = 0$ unless modulo Z the element x is n-admissible of size $(s, r - s)$, where $'\Phi((x', x''), \overline{h}_{n,s}) = s'\Phi(x'', f_0^{(r-s)})$. The existence of $\overline{h}_{n,s}$ is proven in [F2], I.7, using the trace Paley–Wiener theorem of [BDK]. Put $\overline{h}_n = \sum_{s=1}^{r-1} \overline{h}_{n,s}$. The induction assumption implies that $\overline{h}_n$ satisfies $\Phi(\overline{h}_n) = \Phi(\overline{f}_n)$ on the regular nonelliptic set. Hence $\Phi(\overline{f}_n - \overline{h}_n)$ is supported on the elliptic set, and $\mathrm{tr}\,\pi(\overline{f}_n - \overline{h}_n) = 0$ for every nonelliptic G-module π (a G-module π is called *elliptic* if its character is not identically zero on the regular elliptic set).

   (ii) Let π_{P_s} denote the M_s-module of N_s-coinvariants of a G-module π (see [BZ]). The definition (i) of $\overline{h}_n$, the theorem of [C], and the Weyl integration formula imply that $\mathrm{tr}\,\pi(\overline{h}_n) = \sum_{s<r} \mathrm{tr}\,\pi_{P_s}(\overline{h}_{n,s,N_s})$. If $\mathrm{tr}\,\pi(\overline{h}_n) \neq 0$ then there is s such that the M_s-module π_{P_s} contains an irreducible $\pi_s' \times \pi_s''$, where π_s'' is a $\mathrm{GL}(r - s, F)$-module with a nonzero $\mathrm{GL}(r - s, R)$-fixed vector (since $f_0^{(r-s)}$ is spherical), and π_s' is a $\mathrm{GL}(s, F)$-module with a nonzero J-fixed vector (by Corollary 9.8). Suppose that π is irreducible. Then it follows from Frobenius reciprocity that π is a constituent of some unramified induced $I(z)$. Hence π has an Iwahori fixed vector by Lemma 9.5.

(iii) In particular, if π is an irreducible tempered $\mathrm{PGL}(r, F)$-module and $\operatorname{tr} \pi(\overline{f}_n - \overline{h}_n) \neq 0$, then π is elliptic, hence (by [BZ]) ramified (does not have a nonzero K-fixed vector), and $\operatorname{tr} \pi(\overline{h}_n) = \operatorname{tr} \pi(\overline{h}_n - \overline{f}_n) \neq 0$. But then π has (by (ii)) a nonzero J-fixed vector, and by [BZ] the tempered π is of the form $\mathrm{St}_r \otimes \omega_r^i$ for some i.

(iv) Since $\overline{h}_n$ is supported on the x in G with $v(x) \equiv n(\mathrm{mod}\, r)$, we have

$$\operatorname{tr}(\mathrm{St}_r \otimes \omega_r^i)(\overline{h}_n) = \omega_r^i(\pi^n) \operatorname{tr} \mathrm{St}_r(\overline{h}_n) = \zeta_r^{-in} \operatorname{tr} \mathrm{St}_r(\overline{h}_n).$$

(v) By [Z], for each maximal parabolic P_s, we have $(\mathrm{St}_r)_{P_s} = \pi_s' \times \pi_s''$ where π_s', π_s'' are Steinberg $\mathrm{GL}(s, F)$, $\mathrm{GL}(r - s, F)$-modules. In particular π_s'' does not have a $\mathrm{GL}(r - s, R)$-fixed vector unless $r - s = 1$. Hence from now on we take $s = r - 1$ and $P = P_s$. We have

$$\operatorname{tr} \mathrm{St}_r(\overline{h}_n) = \operatorname{tr}(\mathrm{St}_r)_P(\overline{h}_{n,N}) = \operatorname{tr}(\mathrm{St}_r)_P(\overline{h}_{n,s,N}),$$

and

$$(\mathrm{St}_r)_P = (\nu^{1/2} \otimes \mathrm{St}_{r-1}) \times \nu^{(1-r)/2}.$$

Here $\nu(a) = |a|$ is the valuation character on $F^\times$. Note that if $\Phi(x, \overline{h}_n) \neq 0$ then $\nu^{1/2}(\det x') = q^{-n/2}$, which is the inverse of the factor relating $F(f_n^{(r)})$ and $F(f_n^{(s)}) F(f_0^{(r-s)})$; see the proof of Prop. 9.1. Hence

$$\operatorname{tr} \mathrm{St}_r(\overline{h}_n) = (-1)^{s-1} \operatorname{tr} \mathrm{St}_s(u_{n,s}),$$

by Lemma 9.4. This is equal to $(-1)^{s-1}$ times

$$\sum_{i=0}^{s-1} \zeta_s^{ni} \operatorname{tr} \mathrm{St}_s(f_{s,i}) = \operatorname{tr} \mathrm{St}_s(f_{s,0}) = 1.$$

Using (iv) we conclude that

$$\operatorname{tr} \pi\left(\overline{f}_n - \overline{h}_n + (-1)^{s-1} \sum_{i=1}^{r} \zeta_r^{ni} f_{r,i}\right) = 0$$

for all tempered G-modules π. The density theorem of [K1], Appendix (see also [FK1], (19.2), for the special case of $\mathrm{GL}(r)$ which is used here; note that the proof holds also in positive haracteristic), implies that

$$\Phi(x, \overline{f}_n - \overline{h}_n + (-1)^{s-1} u_{n,r}) = 0$$

for all (regular) x. Hence

$${}'\Phi(x, \overline{f}_n) = -(-1)^{s-1}{}'\Phi(x, u_{n,r}) = (-1)^{r-1}{}'\Phi(x, u_{n,r})$$

on the regular elliptic set; there ${}'\Phi(x, u_{n,r})$ is 0, unless $v(x) \equiv n(\mathrm{mod}\, r)$ where $r(-1)^{r-1}$ is obtained.

According to Prop. 9.1, the function $\Phi(x, f_n)$ is supported on the n-admissible x. Hence $\Phi(x, f_n) = 0$ for any nonelliptic regular x sufficiently close to the elliptic set. By the germ expansion of orbital integrals (which follows from the uniqueness of the Haar measure, see [BZ]), ${}'\Phi(x, f_n)$ extends to a continuous function on the entire elliptic set. Since $u_{n,r}$ is a discrete function, the last displayed identity holds for all elliptic x, not necessarily regular. The proposition follows. $\qquad\square$

By the reduction argument explained following Corollary 9.2, we have

Corollary 9.10. *The integral* $\Phi(x, f_n^{(r)})$ *is zero unless x is n-admissible, where* $\Phi(x, f_n^{(r)})$ *is equal to* $e'_v(x)\Phi(x'', \chi''_v)$ *in the notations of Prop.* 8.5.

Proof. This follows from Prop. 9.9 and the relation

$$|Z_x(G)/Z| = [v(Z_x(G)) : v(Z)] = e'(x)/r$$

where $Z_x(G)$ is the centralizer of x in G, and as usual $v(X) = \deg(\det X)$. $\qquad\square$

Denote by $f = f_n^{(r)}$ the spherical function on $G = \mathrm{GL}(r, F)$ with Satake transform

$$f^\vee(z) = q^{n(r-1)/2}(z_1^n + \cdots + z_r^n).$$

Denote by $h = h_n^{(r)}$ the $\mathbb{Z}$-valued spherical function on G which takes the value 0 at $g \in G$ unless $g \in M(r, R) \cap G$ and $v(g) = n$, in which case

$$h(g) = (1 - q)(1 - q^2) \cdots (1 - q^{\dim_k(\ker \bar{g}) - 1});$$

here $\bar{g} \in M(r, k)$, $k = R/m$, is the reduction of g modulo $\pi M(r, R)$, and $\ker \bar{g}$ is the kernel of the endomorphism of k^r defined by $\bar{g}$.

Lemma 9.11. *We have* $f_n^{(r)} = h_n^{(r)}$.

Proof. It suffices to show that $f^\vee = h^\vee$, and this follows by induction on r from the following claim. Denote by $\mathbf{1}_H$ the characteristic function of $H = \mathrm{GL}(t, R)$ in $\mathrm{GL}(t, F)$. Let $P = MN$ be the standard parabolic subgroup of G of type $(1, r-1)$. Then we claim that for any $x = \left(\begin{smallmatrix} a & 0 \\ 0 & b \end{smallmatrix}\right)$ with $a \in F^\times$, $b \in \mathrm{GL}(r-1, F)$, we have

$$(h_n^{(r)})_N(x) = q^{n(r-1)/2}h_n^{(1)}(a)\mathbf{1}_{\mathrm{GL}(r-1,R)}(b) + q^{n/2}h_n^{(r-1)}(b)\mathbf{1}_{R^\times}(a).$$

Note that the unipotent radical N of P consists of the matrices $u = \left(\begin{smallmatrix} 1 & u \\ 0 & 1 \end{smallmatrix}\right)$, $u = (u_1, \ldots, u_{r-1}) \in F^{r-1}$. The modular function is

$$\delta_P(x) = q^{v(b) - (r-1)\deg(a)}.$$

By definition of h the value of $h_n(xu)$ (and so of $(h_n)_N(x)$ too) is zero unless $0 \neq a \in R$, b is a matrix in $\mathrm{GL}(r-1, F)$ with entries in R, and $\deg(a) + v(b) = n$. In this last case

$$(h_n^{(r)})_N(x) = q^{(v(b) - (r-1)\deg(a))/2} \int_{F^{r-1}} h_n^{(r)}(xu) \prod_{1 \leq i < r} du_i.$$

As usual, the measures du_i are normalized to assign R the volume one.

If $a \in R^\times$ and $v(b) = n$, then $h_n^{(r)}(xu)$ is zero unless $u \in R^{r-1}$, in which case $h_n^{(r)}(xu) = h_n^{(r-1)}(b)$, and $h_n^{(1)}(a) = 0$, so the claim follows in this case.

If $\deg(a) = n$ and $b \in \mathrm{GL}(r-1, R)$, then $h_n^{(r)}(xu)$ is zero unless $\boldsymbol{u} \in \boldsymbol{\pi}^{-n} R^{r-1}$, in which case $h_n^{(r)}(xu) = h_n^{(1)}(a)$. Since $h_n^{(r-1)}(b) = 0$, and the volume of $\boldsymbol{\pi}^{-n} R^{r-1}$ is $q^{n(r-1)}$, the claim follows in this case also.

If $\deg(a) > 0$ and $v(b) > 0$, then $h_n^{(r)}(xu)$ is zero unless $\boldsymbol{u} \in a^{-1} R^{r-1}$. In this case $\dim_k \ker(\overline{xu})$ is equal to $\dim_k(\ker \overline{b}) - 1$ if $\overline{au}$ does not lie in the span of the rows of $\overline{b}$. It is equal to $\dim_k(\ker \overline{b})$ if $\overline{au}$ does lie in the span of the rows of $\overline{b}$. Hence, by the definition of $h_n^{(r)}$, we have

$$\int_{F^{r-1}} h_n^{(r)}(xu)\mathrm{d}\boldsymbol{u} = (1-q)\cdots(1-q^{\dim_k(\ker \overline{b})-2})V_1$$

$$+(1-q)\cdots(1-q^{\dim_k(\ker b)-1})V_2.$$

Here

$$V_1 = \mathrm{vol}(\{\boldsymbol{u} \in a^{-1} R^{r-1}; \overline{au} \notin \mathrm{span}\ (\mathrm{rows\ of}\ \overline{b})\})$$

and

$$V_2 = \mathrm{vol}(\{\boldsymbol{u} \in a^{-1} R^{r-1}; \overline{au} \in \mathrm{span}\ (\mathrm{rows\ of}\ \overline{b})\}).$$

The second volume is equal to

$$q^{rk(\overline{b})+(r-1)(\deg(a)-1)}.$$

Hence the first volume is equal to

$$q^{(r-1)\deg(a)} - q^{rk(\overline{b})+(r-1)(\deg(a)-1)},$$

where $rk(\overline{b})$ is the rank of the matrix $\overline{b}$. Consequently $\int_{F^{r-1}} h_n^{(r)}(xu)\mathrm{d}\boldsymbol{u}$ is zero. Since $h_n^{(1)}(a) = 0$ and $h_n^{(r-1)}(b) = 0$ the claim follows in this case too, and the lemma is proven. $\qquad \square$

Lemma 9.11 and Prop. 9.12 are due to Drinfeld (the exposition is influenced by Laumon).

Let γ be an elliptic element of G. Its centralizer $Z(\gamma, G)$ in G is isomorphic to $\mathrm{GL}(r', F')$ where $F' = F[\gamma] \subseteq M(r, F)$ and $r' = r/[F' : F]$. Denote by R' the ring of integers in F' and by q' the cardinality of the residue field k' of R'. Fix the Haar measure $\mathrm{d}g_\gamma$ on $Z(\gamma, G) \simeq \mathrm{GL}(r', F')$ to assign $\mathrm{GL}(r', R')$ the volume one.

Proposition 9.12. *The value* $\Phi(\gamma, f)$ *of the orbital integral of* $f = f_n^{(r)}$ *at the elliptic element* $\gamma \in G$ *is zero unless* $v(\gamma) = n$, *in which case*

$$\Phi(\gamma, f) = (1-q')\cdots(1-q'^{r'-1})[k' : k].$$

Proof. The function f is supported on the set of γ with $v(\gamma) = n$, whence the first assertion. Assuming that $v(\gamma) = n > 0$, and denoting by $\deg' : F' \to \mathbb{Z}$ the discrete valuation on F', we have that $\deg'(\gamma) > 0$. Hence γ lies in the maximal ideal m' of R'. Consequently the minimal polynomial of γ lies in

$R[X]$, and its reduction modulo the maximal ideal $m = \pi R$ of R, is $X^{[F':F]}$. It follows that there exists $g \in G$ with $g^{-1}\gamma g \in M(r, R)$. Choose a set Γ in $\{g \in G'; g^{-1}\gamma g \in M(r, R)\}$ of representatives for the double classes

$$Z(\gamma, G) \backslash \{g \in G; g^{-1}\gamma g \in M(r, R)\}/K, \qquad K = \mathrm{GL}(r, R).$$

Then

$$\Phi(\gamma, f) = \sum_{\eta \in \Gamma} \mathrm{vol}(Z(\gamma, G) \cap \eta K \eta^{-1} \backslash \eta K \eta^{-1}, \, dg/dg_\gamma) f(\eta^{-1}\gamma\eta).$$

Identify G/K with the set of lattices (free rank r R-modules) L in F^r by $gK \mapsto gR^r$. The subset $\{g \in G; g^{-1}\gamma g \in M(r, R)\}/K$ of G/K is mapped to the set of lattices L in F^r with $\gamma L \subset L$. The set Γ is then isomorphic to a set of representatives of the $Z(\gamma, G)$-orbits in the set of lattices L in F^r such that $\gamma L \subset L$. Given $L = \eta R^r$ in Γ we have

$$\eta K \eta^{-1} = \{g \in G; gL = L\}.$$

Moreover

$$\dim_k(\ker(\overline{\eta^{-1}\gamma\eta})) = \dim_k(\ker[\gamma : L/\pi L \to L/\pi L]) = \dim_k(L/(\gamma L + \pi L));$$

we write $d(L)$ for this number. Since the volume of K, and so of $\eta^{-1}K\eta$, is one, one obtains

$$\Phi(\gamma, f) = \sum_{L \in \Gamma} (1 - q) \cdots (1 - q^{d(L)-1})/\mathrm{vol}(\{g \in Z(\gamma, G); gL = L\}, dg_\gamma).$$

Fix an isomorphism of F^r with $F'^{r'}$ as F'-vector spaces. Each lattice L in F^r defines an R'-lattice $R'L$ in $R'^{r'}$. Since Γ consists of representatives of $\mathrm{GL}(r', F')$ orbits, these representatives L can be chosen to satisfy $R'L = R'^{r'}$. Thus Γ will consist of a set of representatives for the $\mathrm{GL}(r', R')$-orbits in the set of lattices L in F^r with $L \subset R'^{r'}$, $R'L = R'^{r'}$, and $\gamma L \subset L$.

Now the reciprocal of the volume of $\{g \in Z(\gamma, G); gL = L\}$ is the index of the subgroup $\{g \in \mathrm{GL}(r', F'); gL = L\}$ in $\mathrm{GL}(r', R')$, since the volume of $\mathrm{GL}(r', R')$ is taken to be one. This is also the number of elements in the $\mathrm{GL}(r', R')$-orbit of L. Hence

$$\Phi(\gamma, f) = \sum_{L} (1 - q) \cdots (1 - q^{d(L)-1}),$$

where L ranges over the set of lattices in F^r with $L \subset R'^{r'}$, $R'L = R'^{r'}$, and $\gamma L \subset L$.

The R-algebra $R[\gamma] \subset R'$ is free of rank $[F' : F]$ as an R-module. It is local with maximal ideal $R[\gamma] \cap m' = (\pi, \gamma)$. The residue field is $R[\gamma]/(\pi, \gamma) = k$.

More generally, let $A \subset R'$ be an R-algebra, free of rank $[F' : F]$ as an R-module, local with maximal ideal $m_A = A \cap m'$, whose residue field A/m_A is k. Put

$$\Phi(A) = \sum_{L} (1 - q)(1 - q^2) \cdots (1 - q^{d(A,L)-1}),$$

where L ranges over the set of lattices L in F^r such that $L \subset R'^{r'}$, $R'L = R'^{r'}$, and $AL \subset L$ and where $d(A, L) = \dim_k(L/m_A L)$. Note that $\Phi(\gamma, f) = \Phi(R[\gamma])$. Moreover the sum over L is finite since for any sufficiently large positive integer t we have $m'^t \subset A$, since an ideal in a local ring contains a power of the maximal ideal, and consequently

$$m'^t R'^{r'} = m'^t R'L = m'^t L \subset AL \subset L \subset R'^{r'}.$$

Let t be the least positive integer with $m'^t \subset A$. We claim that

$$\Phi(A) = (1 - q')(1 - q'^2) \cdots (1 - q'^{r'-1})[k' : k]$$

for all A; the proposition follows once this is proven. The proof is by induction on t. First assume that $t = 1$. Then $m' \subset A$, and $m'R'^{r'} \subset L \subset R'^{r'}$ for each L in the sum which defines $\Phi(A)$. Writing $V = L/m'L \subset k'^{r'}$ one obtains

$$\Phi(A) = \sum_V (1 - q)(1 - q^2) \cdots (1 - q^{\dim_k V - 1}),$$

where V ranges over the finite set of k-vector subspaces of $k'^{r'}$ with $k'V = k'^{r'}$. This is equal to

$$\sum_{W' \neq 0} (-1)^{r'-w'} q'^{(r'-w')(r'-w'-1)/2} D(W'),$$

where W' ranges over the set of k'-vector subspaces in $k'^{r'}$, $w' = \dim_{k'} W'$, and for any k-vector space $W \neq 0$

$$D(W) = \sum_{V \neq 0} (1 - q)(1 - q^2) \cdots (1 - q^{\dim_k V - 1});$$

here V ranges over the set of nonzero k-vector subspaces of W. Indeed, for any k-vector space V

$$\sum_{W'} (-1)^{r'-w'} q'^{(r'-w')(r'-w'-1)/2}$$

is 1 if $k'V = k'^{r'}$ and 0 otherwise, where W' ranges over the set of k'-vector spaces such that $k'V \subset W' \subset k'^{r'}$.

Recall that the number of v-dimensional vector spaces in a w-dimensional vector space is

$$\frac{(1 - q^{w-v+1})(1 - q^{w-v+2}) \cdots (1 - q^w)}{(1 - q)(1 - q^2) \cdots (1 - q^v)}.$$

Consequently, if $w = \dim_k W$, the sum

$$D(W) = \sum_{v=1}^{w} \frac{(1 - q^{w-v+1})(1 - q^{w-v+2}) \cdots (1 - q^w)}{1 - q^v}$$

is equal to $w = \dim_k W$. Hence

$$\Phi(A) = \sum_{W'} (-1)^{r'-w'} w' q'^{(r'-w')(r'-w'-1)/2}[k' : k],$$

where the sum ranges over all nonzero k'-vector subspaces W' of $k'^{r'}$. Note that $w' = \dim_{k'} W'$. Hence $\dim_k W' = w'[k' : k]$. Using again the formula for the cardinality of the Grassmannian, we obtain

$$\Phi(A) = \sum_{w'=1}^{r'} (-1)^{r'-w'} w' q'^{(r'-w')(r'-w'-1)/2} \frac{(1 - q'^{r'-w'+1}) \cdots (1 - q'^{r'})}{(1 - q') \cdots (1 - q'^{w'})} [k' : k]$$
$$= (1 - q')(1 - q'^2) \cdots (1 - q'^{r'-1})[k' : k],$$

as required.

To complete the inductive proof of the claim, if $t \geq 2$ is the least integer with $m'^t \subset A$ (but $m'^{t-1} \not\subset A$), put $A_1 = A + m'^{t-1}$. We proceed to show that $\Phi(A_1) = \Phi(A)$. Note that $\Phi(A_1)$ is well-defined since $A_1 \subset R'$ is an R-subalgebra which is free of rank $[F' : F]$ as an R-module, it is local with maximal ideal $m_{A_1} = m_A + m'^{t-1} = A_1 \cap m'$, and its residue field is equal to that of A, namely to k.

Consider the map $L \mapsto L_1 = A_1 L = AL + m'^{t-1}L$, from the set of lattices L in F^r such that $L \subset R'^{r'}$, $R'L = R'^{r'}$, and $AL \subset L$ to the set of lattices L_1 in F^r such that L_1 lies in $R'^{r'}$, $R'L_1 = R'^{r'}$, and $A_1 L_1 \subset L_1$. This map is clearly surjective. The fiber at L_1 is isomorphic, via the map $L \mapsto V = L/m_A L$, to the set of k-vector subspaces V in $\widetilde{V}_1 = L_1/m_A L_1$, whose image under the surjection $\pi : \widetilde{V}_1 \to V_1 = L_1/m_{A_1} L_1$ is V_1, namely $\pi(V) = V_1$. Consequently

$$\Phi(A) = \sum_{L_1} (1 - q) \cdots (1 - q^{d_1-1}) \varepsilon(\pi : \widetilde{V}_1 \to V_1),$$

where L_1 ranges over the set of lattices in F^r with $L_1 \subset R'^{r'}$, $R'L_1 = R'^{r'}$, and $A_1 L_1 \subset L_1$ and where

$$\varepsilon(\pi : \widetilde{V}_1 \to V_1) = \sum_V (1 - q^{d_1}) \cdots (1 - q^{d-1}).$$

Here V ranges over the set of k-vector subspaces of the $\widetilde{d}_1$-dimensional space $\widetilde{V}_1 = L_1/m_A L_1$ whose image under the surjection π is the d_1-dimensional space $V_1 = L_1/m_{A_1} L_1$. Put d for $\dim_k V$. Then $d_1 \leq d \leq \widetilde{d}_1$.

Put $W_1 = \ker \pi$. The map

$$V \mapsto (W = W_1 \cap V, f : \widetilde{V}_1 \to \widetilde{V}_1/V \xleftarrow{\sim} W_1/W_1 \cap V),$$

from the set of k-vector subspaces V in $\widetilde{V}_1$ with $\pi V = V_1$, to the set of pairs (W, f) where W is a k-vector subspace of W_1 and $f : \widetilde{V}_1 \to W_1/W$ is a k-linear map whose restriction to W_1 is the natural surjection $W_1 \twoheadrightarrow W_1/W$, is a bijection. Hence the number of such V is the number of k-subspaces of dimensional $d - d_1$ in a space of dimension $\widetilde{d}_1 - d_1$, times the number of elements in a k-space of dimension

$$(\dim \widetilde{V}_1 - \dim W_1) \cdot (\dim \widetilde{V}_1 - \dim V) = (\widetilde{d}_1 - (\widetilde{d}_1 - d_1)) \cdot (\widetilde{d}_1 - d) = d_1(\widetilde{d}_1 - d),$$

namely

$$\frac{(1 - q^{\tilde{d}_1 - d + 1}) \cdots (1 - q^{\tilde{d}_1 - d_1})}{(1 - q) \cdots (1 - q^{d - d_1})} q^{d_1(\tilde{d}_1 - d)}.$$

Consequently $\varepsilon(\pi : \tilde{V}_1 \to V_1)$

$$= \sum_{d = d_1}^{\tilde{d}_1} \frac{(1 - q^{d_1}) \cdots (1 - q^{d-1})(1 - q^{\tilde{d}_1 - d + 1}) \cdots (1 - q^{\tilde{d}_1 - d_1})}{(1 - q) \cdots (1 - q^{d - d_1})} q^{d_1(\tilde{d}_1 - d)}$$

is equal to one, and so $\Phi(A) = \Phi(A_1)$, completing the inductive proof of the claim. The proposition follows. $\qquad \Box$

Part 4. Higher Reciprocity Laws

In this part we use the construction of the moduli scheme of elliptic modules, the trace formula, and various forms of the fixed-point formula to obtain various applications. In Chap. 10 we obtain the purity theorem, or Ramanujan conjecture for the cuspidal representations of $GL(r)$ over a function field with a cuspidal component. In Chap. 11 we show that for the cuspidal automorphic representation π with cuspidal component π_∞ there exists a corresponding Galois representation σ. In Chap. 12 we establish the inverse correspondence, thus that the reciprocity law, relating such automorphic representations with irreducible r-dimensional representations of the Weil group as well as a local analogue, is a bijection. To establish the global converse to the existence theorem $\pi \mapsto \sigma$, we develop in Chap. 13 a simple form of the converse theorem, for automorphic representations with a cuspidal component.

10. PURITY THEOREM

The purpose of this chapter is to prove Ramanujan's conjecture for cuspidal representations π of $GL(r, \mathbb{A})$ over a function field, which have a cuspidal component, namely that all unramified components of such a π are tempered, namely that all of their Hecke eigenvalues have absolute value one. This is deduced from a form of the trace formula of Arthur, as well as the theory of elliptic modules developed above, Deligne's purity of the action of the Frobenius on the cohomology, standard unitarity estimates for admissible representations, and Grothendieck's fixed point formula. Once we assume and use Deligne's (proven) conjecture on the fixed point formula, in the following sections, we no longer need the complicated full trace formula, but the simple trace formula suffices. Thus this chapter is for pedagogical purposes only, to show what can be done without Deligne's conjecture.

Recall (Chap. 5) that for any fixed congruences subgroup U_∞ of $D_\infty^\times$, uniformizer π of F_∞, and a congruence subgroup $U = U_I$ in $G(\widehat{\mathbb{A}})$, we have a finite étale Galois covering $\widetilde{M}_{r,I} = U \backslash \widetilde{M}_r = \operatorname{Spec} \widetilde{B}^U$ of $M_{r,I} = U \backslash M_r = \operatorname{Spec} B^U$ with Galois group $\Gamma = \langle \pi \rangle U_\infty \backslash D_\infty^\times$. Fix v in $\operatorname{Spec} A$ and put $\mathbb{F}_v = A/v$. Let $\widetilde{X}_I = \widetilde{M}_{r,v,I}(\overline{\mathbb{F}}_p)$ be the set of $\overline{\mathbb{F}}_p$-points on the fiber $\widetilde{M}_{r,v,I} = \widetilde{M}_{r,I} \times_{\operatorname{Spec} A} \operatorname{Spec} \mathbb{F}_v$ of $\widetilde{M}_{r,I}$ at v. The Galois group Γ and the adèle group $G(\mathbb{A}_f)$ act on the set $\varprojlim \widetilde{X}_J$. These two actions commute. The group $F^\times$, embedded diagonally in the direct product $\Gamma \times G(\mathbb{A}_f)$, acts trivially. Moreover, $\widetilde{X}_I = U \backslash \varprojlim \widetilde{X}_J$. By Corollary 7.5 the set $\widetilde{X} = \widetilde{X}_I$ decomposes as the disjoint union of sets $\widetilde{Y}$ parametrized by (F, v)-types (F', v'), which by Prop. 8.1 are of the form

$$\widetilde{Y} = (\Gamma \times U \backslash G(\mathbb{A}_f)/S_v N_v)/D'^\times;$$

recall that $D'^\times$ naturally embeds in $(D' \otimes_F F_\infty)^\times = D_\infty^\times$. The set $X = X_I = M_{r,v,I}(\overline{\mathbb{F}}_p)$ is the disjoint union over the types (F', v') of the sets

$$Y = (U \backslash G(\mathbb{A}_f)/S_v N_v)/D'^\times.$$

Let S, $\widetilde{S}$ be sets, $h : \widetilde{S} \to S$ an epimorphism, and Γ a finite group. Suppose that Γ acts simply transitively on $h^{-1}(s)$ for every s in S. Let $\widetilde{A}$ be an automorphism of $\widetilde{S}$ and A of S such that $Ah = h\widetilde{A}$ and $g\widetilde{A} = \widetilde{A}g$ for every g in Γ. Given s in S choose $\widetilde{s}$ in $h^{-1}(s)$ and define g in Γ by $\widetilde{A}\widetilde{s} = g\widetilde{s}$. Let $\beta(s)$ be the conjugacy class of g.

Lemma 10.1. *The map $s \mapsto \beta(s)$ is a well-defined map from the set S^A of A-fixed points in S to the set $X(\Gamma)$ of conjugacy classes in Γ.*

Proof. Let $\widetilde{s}'$ be any element of $\widetilde{S}$ with $h(\widetilde{s}') = s$. Define g' in Γ by $\widetilde{A}\widetilde{s}' = g'\widetilde{s}'$. For some x in Γ we have $\widetilde{s}' = x\widetilde{s}$. Then $g'x\widetilde{s} = g'\widetilde{s}' = \widetilde{A}\widetilde{s}' = \widetilde{A}x\widetilde{s} = x\widetilde{A}\widetilde{s} = xg\widetilde{s}$, and g, g' define the same conjugacy class. $\square$

In our case we put $\mathfrak{F}^n = 1 \times \operatorname{Fr}_v^n$. It acts on X and on $\widetilde{X}$. It commutes with the action of Γ. Lemma 10.1 defines a map $x \mapsto \gamma_\infty$ from the set $X^{\mathfrak{F}^n}$

Y.Z. Flicker, *Drinfeld Moduli Schemes and Automorphic Forms: The Theory of Elliptic Modules with Applications*, SpringerBriefs in Mathematics, DOI 10.1007/978-1-4614-5888-3_10, © Yuval Z. Flicker 2013

of $\mathfrak{F}^n$-fixed points in X, to the set of conjugacy classes in Γ. If $\widetilde{x}$ in $\widetilde{Y}$ is represented by $g = (g_\infty, z, g_v'', g^v)$ with g_∞ in Γ and $\mathfrak{F}\widetilde{x} = \gamma_\infty \widetilde{x}$, then

$$(g_\infty, z+n, g_v'', g^v) = (\gamma_\infty g_\infty \gamma, z + v(\gamma), (u_v'')^{-1} g_v'' \gamma, (u^v)^{-1} g^v \gamma)$$

for some γ in $D'^\times$ and $u = u_v'' \times u^v$ in U. Hence $\gamma_\infty = g_\infty \gamma^{-1} g_\infty^{-1}$, and the image x in Y of $\widetilde{x}$ is mapped to the conjugacy class of γ^{-1} in Γ. We write $\gamma(x)$ for γ.

Let (ρ, V) be an irreducible $D_\infty^\times$-module whose central character ω has finite order. Then there is a congruence subgroup U_∞ of $D_\infty^\times$ such that ρ is trivial on U_∞. Multiplying ρ by an unramified character we may assume that the value at π of the central character of ρ is one. Hence we can view ρ as a representation of the finite group Γ. Denote by $\operatorname{tr} \rho(\gamma)$ the character of ρ at γ. The values taken by $\operatorname{tr} \rho$ are clearly algebraic.

We shall use the Grothendieck fixed point formula (Theorem 6.6) in the ℓ-adic cohomology of the scheme $\overline{X} = \overline{M}_{r,v,I} = M_{r,v,I} \times_{\mathbb{F}_v} \overline{\mathbb{F}}_v$, where $M_{r,v,I}$ is an affine, hence separated scheme of finite type over the finite field $\mathbb{F}_v$, and $\overline{\mathbb{F}}_v$ is an algebraic closure of $\mathbb{F}_v$. The coefficients are taken in the smooth $\overline{\mathbb{Q}}_\ell$-sheaf $\mathbb{L}(\rho)$ determined (see Sect. 6.1.4) by the ℓ-adic representation (ρ, V), where $V = \overline{\mathbb{Q}}_\ell^{n_\rho}$, of the finite Galois group Γ. Thus let $H_\rho^* = H_\rho^*(v, U, U_\infty)$ be the alternating sum $\sum_i (-1)^i H_c^i(\overline{X}, \mathbb{L}(\rho))$ of the $\overline{\mathbb{Q}}_\ell$-adic cohomology spaces of $\overline{X}$ with compact support and coefficients in the smooth $\overline{\mathbb{Q}}_\ell$-sheaf $\mathbb{L}(\rho)$. Both the Hecke algebra $\mathbb{H}_I$, which is generated by the characteristic functions of the double cosets UgU (g in $G(\mathbb{A}_f)$), and the Frobenius $1 \times \operatorname{Fr}_v$ act on the cohomology H_ρ^* (as explained in Sect. 6.1). The Galois group Γ acts on V via ρ, hence also on H_ρ^*. Hence H_ρ^* is a virtual $(\Gamma \times \mathbb{H}_I)/F^\times$- and $\langle 1 \times \operatorname{Fr}_v \rangle$-module over $\overline{\mathbb{Q}}_\ell$. As in Chap. 6 we denote by $\operatorname{Fr}_v \times 1$ the geometric Frobenius, which acts on the cohomology as the inverse of the arithmetic Frobenius $1 \times \operatorname{Fr}_v$ (see Sect. 6.1.9). We put again $\mathfrak{F}^n = 1 \times \operatorname{Fr}_v^n$ and also $\overline{\mathfrak{F}}^n = \operatorname{Fr}_v^n \times 1$. The Grothendieck fixed point formula (Theorem 6.6) asserts

Lemma 10.2. *For any integer $n \neq 0$ we have*

$$\operatorname{tr}[\overline{\mathfrak{F}}^n | H_\rho^*] = \sum_{x \in X^{\overline{\mathfrak{F}}^n}} \operatorname{tr} \rho(\gamma(x)^{-1}).$$

Denote the center of G_∞ by Z_∞; it is isomorphic to the center of $D_\infty^\times$ and to $F_\infty^\times$. Recall that the central character ω of the irreducible ρ is assumed to be of finite order, in particular unitary. Let $\pi_\infty(\rho)$ be the square-integrable G_∞-module which corresponds (see, e.g., [F2], [III]) to the $D_\infty^\times$-module ρ. Let f_∞ be a locally constant complex-valued function on G_∞ with $f_\infty(zx) = \omega(z)^{-1} f_\infty(x)$ for all x in G_∞, z in Z_∞, which is compactly supported on G_∞ modulo Z_∞. We require that at each regular x in G_∞ the orbital integral $\Phi(x, f_\infty)$ be zero, unless x is elliptic where $'\Phi(x, f_\infty) = (-1)^{r-1} \operatorname{tr} \rho(x^{-1})$. Then f_∞ is a pseudo-coefficient (see [K1] and Definition 9.1) of the irreducible G_∞-module $\pi_\infty(\rho)$. Below we take $\pi_\infty(\rho)$ to be cuspidal, in which case f_∞ can be taken to be a normalized matrix coefficient of $\pi_\infty(\rho)$, which is compactly supported modulo Z_∞.

The computations of the structure of the set $X^{\overline{\mathfrak{F}}^n}$ of fixed points of the geometric Frobenius $\overline{\mathfrak{F}}^n$ in X yield

Proposition 10.3. *For any $n \neq 0$ the trace* $\mathrm{tr}[\overline{\mathfrak{F}}^n | H_\rho^*]$ *is equal to*

$$(-1)^{r-1} \sum_\gamma |Z_\gamma(G(\mathbb{A}_f))/Z_\gamma(G(F))| \cdot {}'\Phi(\gamma, f_\infty)\Phi(\gamma, f_{n,v})\Phi(\gamma, \chi^v)$$

$$= (-1)^{r-1} \sum_\gamma |Z_\gamma(G(\mathbb{A}))/Z_\gamma(G(F))Z_\infty| \cdot \Phi(\gamma, f_\infty f_{n,v}\chi^v).$$

The sums range over all elliptic conjugacy classes γ in $G(F)$; $Z_\gamma(G)$ is the centralizer of γ in G.

Proof. The first equality follows from Lemma 10.1, Prop. 8.5, and Corollary 9.10. The second equality follows from the relation

$${}'\Phi(\gamma, f_\infty) = |Z_\gamma(G_\infty)/Z_\infty|\Phi(\gamma, f_\infty)$$

for an elliptic γ in G_∞. Note that if γ contributes a nonzero term to the first sum, then it is elliptic in G_∞, hence in $G(F)$, and n-admissible in G_v. $\quad\square$

For use in Chap. 11 we record here a variant. Let f^∞ be a compactly supported locally constant $\overline{\mathbb{Q}}_\ell$-valued function on $G(\mathbb{A}_f)$. It defines a correspondence on $\overline{X}$ and an automorphism of $H_c^i(\overline{X}, \mathbb{L}(\rho))$ as in Sect. 6.1.8, denoted here again by f^∞.

Proposition 10.4. *There exists a positive integer $n_0 = n_0(f^\infty)$ such that for every $n \geq n_0$ we have*

$$\mathrm{tr}[f^\infty \cdot \overline{\mathfrak{F}}^n | H_\rho^*] = (-1)^{r-1} \sum_\gamma |Z_\gamma(G(\mathbb{A}))/Z_\gamma(G(F))Z_\infty|\Phi(\gamma, f_\infty f_{n,v}f^\infty).$$
(10.1)

The sum ranges over the set of elliptic conjugacy classes γ in $G(F)$.

Proof. This follows from Deligne's conjecture (Theorem 6.8), Prop. 8.5, and Corollary 9.10. $\quad\square$

Remark 10.1. We fix an embedding and an isomorphism $\overline{\mathbb{Q}} \hookrightarrow \overline{\mathbb{Q}}_\ell \simeq \mathbb{C}$ and regard the right sides of the formulae in Lemma 10.2, Prop. 10.3, and Prop. 10.4 as complex numbers.

Our next aim is to show that the sum of Prop. 10.3 appears as one of the sides in the Selberg trace formula of Prop. 10.6. Let F be a global function field. Fix a character ω of finite order of the center Z_∞ of G_∞, where ∞ denotes a fixed place of F. At each place w of F choose the Haar measure dg_w on $G_w = G(F_w)$ which assigns $G(R_w)$ the volume one. Denote by dg the product measure $\otimes dg_w$ on $G(\mathbb{A})$. Put $(r(g)\varphi)(h) = \varphi(hg)$ $(h, g$ in $G(\mathbb{A}))$ for a function φ on $G(\mathbb{A})$ and $(r(f_v)\varphi)(h) = \int_{G_v} f_v(g)\varphi(hg)dg$ for any f_v in the algebra $\mathbb{H}_v$ of compactly supported $\overline{\mathbb{Q}}_\ell$-valued locally constant functions f_v on G_v.

Let $L^2(G)$ be the span of the complex-valued functions φ on $G(F)\backslash G(\mathbb{A})$ with $\varphi(zx) = \omega(z)\varphi(x)$ $(z$ in Z_∞, x in $G(\mathbb{A}))$ which are absolutely square integrable on $Z_\infty G(F)\backslash G(\mathbb{A})$ and are eigenvectors of $r(\mathbb{H}_v)$ for almost all v. Then $r(G(\mathbb{A}))\varphi$ is an admissible $G(\mathbb{A})$-module for each φ in $L^2(G)$. Each irreducible constituent of the $(G(\mathbb{A}), r)$-module $L^2(G)$ is called an *automorphic* $G(\mathbb{A})$-module, or automorphic representation.

A cuspidal function φ on $G(F)\backslash G(\mathbb{A})$ is one satisfying $\int_{N(F)\backslash N(\mathbb{A})} \varphi(nx)dn = 0$ for every x in $G(\mathbb{A})$ and every proper F-parabolic subgroup $P = MN$ of G with unipotent radical N. The space $L_0(G)$ of cusp forms (= functions) decomposes as a direct sum with finite multiplicities of irreducible $G(\mathbb{A})$-modules, called *cuspidal* $G(\mathbb{A})$-modules. These multiplicities are equal to one by the "multiplicity one theorem" for $\mathrm{GL}(r)$.

Suppose that $\pi_\infty(\rho)$ is a *cuspidal* G_∞-module with central character ω. Thus for each matrix coefficient f_∞ of $\pi_\infty(\rho)$ and each proper parabolic subgroup P of G_∞ over F_∞, we have $\int_N f_\infty(xny)dn = 0$ for all x, y in G_∞, where N denotes the unipotent radical of P. Let $L_\rho^2(G)$ denote the subspace of $L^2(G)$ where $G(\mathbb{A})$ acts as a multiple of $\pi_\infty(\rho)$. It is well known that $L_\rho^2(G)$ is a subspace of the space of cusp forms. Hence it decomposes as a direct sum of inequivalent irreducible $G(\mathbb{A})$-modules $\pi = \otimes\pi_w$ with $\pi_\infty = \pi_\infty(\rho)$.

Let $f = \otimes f_w$ be a complex-valued locally constant function on $G(\mathbb{A})$ with the following properties. We take f_∞ to be a normalized matrix coefficient of $\pi_\infty(\rho)$. Hence f_∞ transforms by ω^{-1} on Z_∞, it is compactly supported modulo Z_∞, and for every irreducible G_∞-module π'_∞ with central character ω, $\mathrm{tr}\,\pi'_\infty(f_\infty)$ is 0 unless π'_∞ is $\pi_\infty(\rho)$, where $\mathrm{tr}\,\pi'_\infty(f_\infty) = 1$. At each $w \neq \infty$ the component f_w is taken to be compactly supported. At almost all places w it is taken to be the characteristic function of $G(R_w)$. The convolution operator $r(f)$ acts on $L_\rho^2(G)$; it is an integral operator with kernel $K_f(x, y) = \sum_\gamma f(x\gamma y^{-1})$. The sum ranges over all γ in $G(F)$. The operator $r(f)$ is of trace class. Its trace is given by $\int_{G(\mathbb{A})/G(F)Z_\infty} K_f(x, x)dx$.

Lemma 10.5. *The set of conjugacy classes γ in $G(F)$ for which there exists x in $G(\mathbb{A})$ with $f(x\gamma x^{-1}) \neq 0$ is finite; it depends only on the support of f.*

Proof. The map sending γ to the ordered set $(a_1, \ldots, a_r)$ of coefficients in the characteristic polynomial of γ is a bijection from the set of semisimple conjugacy classes in $G(\mathbb{A})/Z_\infty$ to the quotient of $\mathbb{A}^{r-1} \times \mathbb{A}^\times$ by the relation $(a_1, \ldots, a_r) \equiv (a_1 z, \ldots, a_r z^r)$ (z in $F_\infty^\times$). The image of $G(F)$ is discrete. The image of the support of f is compact. There are only finitely many conjugacy classes with the same semisimple part. $\square$

The following is the trace formula.

Proposition 10.6. *Put $f = f_\infty f_v f^{v,\infty}$. Suppose that f_v is the spherical function $f_{n,v}$ of Definition 9.1 and that n is sufficiently large with respect to (the support of) f_∞ and $f^{v,\infty} = \otimes_{w \neq v,\infty} f_w$. Then the trace $\mathrm{tr}\,r(f)$ of $r(f)$ on $L_\rho^2(G)$ is equal to*

$$\sum_\pi \mathrm{tr}\,\pi(f) = \sum_\gamma c(\gamma)\Phi(\gamma, f),$$

where

$$c(\gamma) = |Z_\gamma(G(\mathbb{A}))/Z_\infty Z_\gamma(G)|.$$

On the left the sum ranges over all irreducible $G(\mathbb{A})$-modules π in $L_\rho^2(G)$.
On the right the sum ranges over all elliptic conjugacy classes γ in $G(F)$, such
that γ is n-admissible in G_v and elliptic in G_∞.

Proof.   (i) Since $r(f)$ is a trace class operator on $L_\rho^2(G) = \oplus \pi$ and each
π occurs with multiplicity one in this sum, it is clear that its trace
is given by the left side of the identity of the proposition. We need
to prove the identity of the proposition. Both sides of the proposed
identity depend only on the orbital integral $\Phi(x, f_v)$ of f_v at x. Indeed,
this is clear for the right side. For the left side, suppose that $\Phi(x, f_v)$
is zero for all x in G_v. Then, using the uniqueness of the Haar measure
[BZ], standard properties of closure of orbits (recorded, e.g., in [F2],
end of p. 160), and the compactness of $\operatorname{supp} f_v$, one concludes (cf. [BZ],
Theorems 6.9/10, p. 54/55) that there are finitely many h_i in $C_c(G_v)$
and g_i in G_v, with $f_v = \sum_i (h_i^{g_i} - h_i)$ on G_v. Hence $\operatorname{tr} \pi_v(f_v)$ vanishes
if $\Phi(x, f_v)$ is zero for all x. Consequently we may replace f_v by any
other function which has the same orbital integrals.

(ii) By virtue of Corollary 9.2 it suffices to prove the identity of the proposition where the component f_v is replaced by (1) any discrete function
on G_v and (2) a function on G_v which vanishes on the elliptic set whose
orbital integrals are equal to those of $f_{n,v}$ on the nonelliptic set.

(iii) To deal with the first case, suppose that $f = f_\infty f_v f^{v,\infty}$, where f_v
is discrete. Denote by G' the multiplicative group of the division
algebra D' of dimension r^2 central over F, which is split at each place
$w \neq v, \infty$ and is defined by $\operatorname{inv}_v D' = 1/r$ (and $\operatorname{inv}_\infty D' = -1/r$).
Then $G'_w = G'(F_w)$ is isomorphic to $G_w = G(F_w)$ for all $w \neq v, \infty$, and
G'_v (resp. G'_∞) is an anisotropic inner form of G_v (resp. G_∞). To prove
the identity we use the correspondence from G' to G; see, e.g., [F2],
III, when the characteristic is zero, and note that the analogous results
can be transferred to the case of positive characteristic on using [K2].

Remark 10.2. Note that the problem in establishing the identity of Prop. 10.6
in our case is the evaluation of the contributions parametrized by the singular γ
on the right. In characteristic zero this follows from the explicit computations
of some sequel of [A]. The proof here, in the context of $\operatorname{GL}(r)$ only, is valid
in all characteristics and uses the correspondence to deduce the identity of
Prop. 10.6 from the trace formula for the anisotropic group G'.

We first recall that there is a bijection from the set of conjugacy classes γ'
in G'_v (resp. G'_∞) to the set of elliptic conjugacy classes γ in G_v (resp. G_∞):
γ' and γ correspond if γ' and γ have the same characteristic polynomials. Also
there is a bijection, defined in the same way, from the set of conjugacy classes
in $G'(F)$ to the set of elliptic conjugacy classes in $G(F)$ which are elliptic at
v and ∞.

Next we recall the definition of transfer of functions from G_w to G'_w. If
$w \neq v, \infty$, then $G'_w \simeq G_w$ and f_w on G_w defines a function f'_w on G'_w via this
isomorphism. The locally constant compactly supported functions f'_v on G'_v
and f_v on G_v are called *matching* if f_v is discrete (its orbital integrals vanish on

the regular nonelliptic set), and $\Phi(\gamma, f_v) = \Phi(\gamma', f'_v)$ for every regular elliptic γ in G_v (γ' indicates the corresponding class in G'_v). A basic result asserts that for every f'_v (resp. discrete f_v) there exists a matching f_v (resp. f'_v), and that $\Phi(\gamma', f'_v) = (-1)^{r_v(\gamma)}\Phi(\gamma, f_v)$ for every pair (γ, γ') of matching elements. Here $r_v(\gamma) = \dim(A_\gamma/Z_v)$, where A_γ is a maximal split torus in the centralizer $Z_\gamma(G_v)$ of γ in G_v (in particular $r_v(\gamma) = 0$ if γ is elliptic regular). An analogous definition is introduced also for locally constant compactly supported modulo Z_∞ functions f'_∞ on G'_∞ and f_∞ on G_∞.

Finally we recall that the correspondence is a bijection from the set of equivalence classes of irreducible G'_v-modules π'_v to the set of equivalence classes of square-integrable G_v-modules π_v. It is defined by $\operatorname{tr}\pi_v(f_v) = \operatorname{tr}\pi'_v(f'_v)$ for all matching (f_v, f'_v). Similarly we have the correspondence $\pi'_\infty \mapsto \pi_\infty$. Fix a G'_∞-module ρ such that the corresponding G_∞-module $\pi_\infty(\rho)$ is cuspidal. Globally there is (see, e.g., [F6] or [F2], III) a bijection from the set of cuspidal $G'(\mathbb{A})$-modules $\pi' = \otimes\pi'_w$ with $\pi'_\infty = \rho$ to the set of cuspidal $G(\mathbb{A})$-modules $\pi = \otimes\pi_w$ with $\pi_\infty = \pi_\infty(\rho)$ such that π_v is square integrable. It is defined by $\pi_w \simeq \pi'_w$ for all $w \neq v, \infty$, and $\pi'_v \mapsto \pi_v$ at v. In particular for corresponding global functions $f = \otimes f_w$ and $f' = \otimes f'_w$ we have

$$\sum_\pi \operatorname{tr}\pi(f) = \sum_{\pi'} \operatorname{tr}\pi'(f').$$

The first sum ranges over all cuspidal $G(\mathbb{A})$-modules whose component at ∞ is the cuspidal $\pi_\infty(\rho)$ (and its component at v is necessarily square integrable). The second sum ranges over all cuspidal $G'(\mathbb{A})$-modules π' whose component at ∞ is ρ.

We are now ready to prove the identity of the proposition for our function f. The trace formula for the anisotropic group $G'(\mathbb{A})$ and the function f' asserts that

$$\sum_{\pi'} \operatorname{tr}\pi'(f') = \sum_{\gamma'} c(\gamma')\Phi(\gamma', f'),$$

where

$$c(\gamma') = |Z_{\gamma'}(G'(\mathbb{A}))/Z_{\gamma'}(G'(F))Z_\infty|.$$

The sum over π' is the same as in the identity of the proposition. The sum on the right ranges over all conjugacy classes γ' in $G'(F)$. Since $f' = \otimes f'_w$ matches $f = \otimes f_w$, the sum on the right of the trace formula is equal to the sum on the right of the identity of the proposition. Indeed, $\Phi(\gamma, f)$ is zero unless γ is elliptic and corresponds to some γ' in the trace formula, in which case $\Phi(\gamma, f) = \Phi(\gamma', f')$. The volume factors $c(\gamma')$ and $c(\gamma)$ are equal since $Z_{\gamma'}(G')$ is an inner form of $Z_\gamma(G)$. Hence the two groups have equal Tamagawa numbers. This completes the proof of the identity of the proposition in the case that the component f_v of f is discrete.

(iv) It remains to prove the identity of the proposition where f_v vanishes on the elliptic set and its orbital integral is equal to that of $f_{n,v}$ on the nonelliptic set, where n is sufficiently large. We first note that the set S of x in G_v whose semisimple part is n-admissible but not elliptic is open and closed in G_v. Since the orbital integral of f_v is supported on S, by virtue of (i) we may replace f_v by its product with the characteristic function of S. Namely we may assume that f_v is supported on S. In particular, if $f(x\gamma x^{-1}) \neq 0$ for some x in $G(\mathbb{A})$ then the semisimple part γ' of γ is n-admissible, but not elliptic, in G_v.

(v) Suppose that γ is an element of $G(F)$ such that $f(x\gamma x^{-1}) \neq 0$ for some x in $G(\mathbb{A})$. We shall show that if n is sufficiently large (depending on f_w ($w \neq v$)), then the semisimple part γ' of γ is elliptic in $G(F)$. Indeed, we have $Z_{\gamma'}(F) = \prod_i \mathrm{GL}(r_i, F_i)$, $1 \leq i \leq t$, with $\sum_i r_i[F_i : F] = r$. Write $\gamma' = (\gamma_1, \ldots, \gamma_t)$ with γ_i in $\mathrm{GL}(r_i, F_i)$, correspondingly. Put $x_{iw} = \deg_w(N_{F_i/F}\gamma_i)$ for each valuation w of F. Since the f_w ($w \neq v$) are fixed, for every i, j, and $w \neq v$ the difference $x_{iw} - x_{jw}$ lies in a finite set (of integers). This difference is equal to zero for almost all w (depending on $\otimes_{w \neq v} f_w$), and by the product formula on F we have that $\sum_w (x_{iw} - x_{jw}) = 0$ for all i and j. Hence $x_{iv} - x_{jv}$ lies in a fixed finite set for all i and j. The choice of f_v in (iv) guarantees that γ' is n-admissible, hence that there is some i for which x_{iv} attains the value n, while x_{jv} attains the value zero for all $j \neq i$. Consequently, if n is sufficiently large and $f(x\gamma x^{-1}) \neq 0$ for some x in $G(\mathbb{A})$, then γ' is elliptic in G (i.e., $t = 1$).

(vi) We shall prove the identity of the proposition for f with f_v as in (iv) on using the computation of

$$\int_{G(\mathbb{A})/GZ_\infty} \left[\sum_{\gamma \in G} f(x\gamma x^{-1}) \right] \mathrm{d}x$$

in [A]. The theorems of [A] are stated only for number fields. We shall use them in our function field case or, alternatively, complete the proof of the number field analogue of the identity of the proposition. Our proof then depends on verifying that the statements of [A] hold in the positive characteristic case. It seems that this can be done on making only minor changes to the techniques of [A]. However, this we do not do here.

Note that our definition of a semisimple element as before Definition 8.1 coincides with that of [A1], top half of p. 921. In characteristic zero, a semisimple element as defined as before Definition 8.1 is necessarily diagonalizable. This is the usual definition of semisimplicity, which is implicitly recorded in [A1], end of p. 920. However, only the properties recorded in [A1], p. 921, are used in the work of [A]. They are the ones used to define semisimplicity in our case of positive characteristic, as before Definition 8.1.

Other changes are as follows: the exponential function e of [A1], p. 945, should be replaced by $e(x) = 1 + x$; the positive definite bilinear form $\langle ., . \rangle$ by a nondegenerate bilinear form; ψ by a nontrivial character on $\mathbb{A}/F$; and the lattice $\mathbb{Z}$ in $\mathbb{R}$ (on p. 946) by a lattice in F_∞. However, as noted above, we shall use, but not prove here, the positive characteristic analogue of [A].

As explained in Chap. 11, using Deligne's conjecture permits us not to use the work of [A], which we use here to establish Ramanujan's conjecture without using Deligne's conjecture.

Let $\mathcal{O}$ be the set of equivalence classes of γ in G, where γ and γ' are said to be equivalent if their semisimple parts are conjugate (in G). Put $J_\vartheta(f) = \int k_\vartheta(x, f)\mathrm{d}x$, where $k_\vartheta(x, f) = K_{G,\vartheta}(x, x)$ is defined to be $\sum_{\delta \in \vartheta} f(x^{-1}\delta x)$, for any $\vartheta \in \mathcal{O}$. Since f_∞ is a cusp form, [A1], Theorem 7.1, p. 942, asserts the convergence of the sum $\sum_{\vartheta \in \mathcal{O}} \int |k_\vartheta(x, f)|\mathrm{d}x$ (the cuspidality of f_∞ easily implies the vanishing of the terms $K_{P,\vartheta}(x, x)$, $P \neq G$, which appear in the definition of $k_\vartheta^T(x, f)$ in [A1], p. 938, and the independence of $k_\vartheta^T(x, f)$ of the auxiliary parameter T). To compute the $J_\vartheta(f)$ we shall use the formal, geometric computations of [A2,3].

Denote by $\gamma_\vartheta = (\gamma_1, \ldots, \gamma_t)$ ($\in Z_{\gamma_\vartheta}(F)$) a semisimple element in ϑ, in the notations of (v). If $t \neq 1$, then the argument of (v) shows that for a sufficiently large n the kernel $K_\vartheta(x, x) = K_{G,\vartheta}(x, x)$ is identically zero. Thus we may assume that γ_ϑ is elliptic, namely $t = 1$. For a general ϑ the term $J_\vartheta(f)$ is expressed in [A2], Theorem 8.1, p. 206, as a linear combination of functions $J_M(\gamma', f)$, $\gamma' \in \vartheta \cap M$, which are defined in [A3], (6.5), p. 254, and [A3], Theorem 5.2, p. 245. Here M is a standard Levi subgroup of G; denote its center by A_M. Fix a finite set S of places of F such that $f_w = f_w^0$ for all w outside S. The definition of $J_M(\gamma', f)$ involves a limit over $a \to 1$, $a \in A_M(F_S) = \prod_{w \in S} A_M(F_w)$, of "corrected" weighted orbital integrals (see [A3], (2.1), p. 234):

$$J_L(a\gamma', f) = |D(a\gamma')|^{1/2} \int_{Z_\infty G_{a\gamma'}(F_S) \backslash G(F_S)} f(x^{-1}a\gamma'x)v_L(x)\mathrm{d}x.$$

Here L is an element in the set $\mathcal{L}(M)$ of Levi subgroups of G which contain M ([A3], p. 228). Since $a\gamma'$ is regular for a generic $a \in A_M(F_S)$, the argument of (v) shows that for a large n the integral $J_L(a\gamma', f)$ is zero unless $M = G$. Then [A2], (8.2), asserts that $J_\vartheta(f)$ is a linear combination over the conjugacy classes γ' in $\vartheta \cap G(F_S)$ of the orbital integrals $J_G(\gamma', f)$ of f at γ'. If such γ' is not semisimple then the orbital integral $J_G(\gamma', f_\infty)$ vanishes since f_∞ is a cusp form. Then [A2], (8.2), implies that $J_\vartheta(f) = a^G(S, \gamma)J_G(\gamma, f)$, where $\gamma = \gamma_\vartheta$ is elliptic (semisimple) in ϑ. But [A2], (8.1), p. 206, implies that $a^G(S, \gamma) = a^{G_\gamma}(S, 1) = |Z_\infty G_\gamma \backslash G_\gamma(\mathbb{A})| = c(\gamma)$ in our notations, and the identity of the proposition follows.   □

Remark 10.3. The "corrected" weighted orbital integrals were first introduced in [F1] in the context of GL(3) (and GL(2)), where $\sum_\vartheta J_\vartheta^T(f)$ is explicitly computed and related to the limit values of the corrected weighted orbital integrals on regular classes. In (vi) above we use the generalization to GL(n) of [A2,3].

Corollary 10.7. *Suppose that F is a function field, $f^{v,\infty}$ is χ^v, f_∞ is a normalized matrix coefficient of the cuspidal representation $\pi_\infty(\rho)$, and n is large with respect to $f^{v,\infty}$ and f_∞. Then*

$$\sum_i (-1)^{r-1+i} \operatorname{tr}[(\operatorname{Fr}_v^n \times 1)|H_c^i(\overline{X}, \mathbb{L}(\rho))]$$

$$= \sum_\pi \operatorname{tr} \pi(f) = \sum_\pi n(\pi, U) q_v^{n(r-1)/2} \sum_i z_i(\pi_v)^n.$$

The sum over π is finite. It ranges over all irreducible constituents of $L_\rho^2(G)$ such that $\pi^\infty = \otimes_{w \neq \infty} \pi_w$ has a nonzero vector fixed under the action of $U = U_I$. Further, $n(\pi, U)$ is the integer $\prod_{w \neq v, \infty} \operatorname{tr} \pi_w(f_w)$.

Proof. The first equality results from the expressions for $\operatorname{tr}[\overline{\mathfrak{F}}^n|H_\rho^*]$ and $\sum_\pi \operatorname{tr} \pi(f)$ in terms of $G(\mathbb{A})$ orbital integrals on the group $G(F)$. The conditions on n guarantee that we deal with elliptic conjugacy classes only. Since $f_{n,v}$ is spherical we have that $\pi_v = \pi_v((z_i(\pi_v)))$ if $\operatorname{tr} \pi_v(f_{n,v}) \neq 0$. But then $\operatorname{tr}(\pi_v(z))(f_{n,v}) = q_v^{n(r-1)/2} \sum_i z_i(\pi_v)^n$ by the definition of $f_{n,v}$. Note that there are only finitely many cuspidal $G(\mathbb{A})$-modules with fixed ramification at all places and in particular π in $L_\rho^2(G)$ with $n(\pi, U) \neq 0$. The corollary follows. $\qquad\square$

We shall now use Corollary 10.7 to prove the following purity theorem, or Ramanujan's conjecture, for certain cusp forms of $\operatorname{GL}(r, F)$.

Theorem 10.8. *Let $\pi = \otimes \pi_w$ be an automorphic $G(\mathbb{A})$-module whose central character is unitary and whose component π_∞ at ∞ is cuspidal. Then for every place v where π_v is unramified we have that each Hecke eigenvalue $z_{i,v} = z_i(\pi_v)$ $(1 \leq i \leq r)$ of π has complex absolute value equal to one.*

Proof. We first tensor π with an everywhere unramified character to assure that the central character ω of π, which we now denote by π', is of finite order. Let U be a sufficiently small congruence subgroup of $G(\mathbb{A}_f)$ (the place ∞ is fixed in the statement of the theorem), with component $U_v = \operatorname{GL}(r, R_v)$ at the place v, such that π' has a nonzero U-fixed vector. Then π' appears in the set over which the sum $\sum_\pi \operatorname{tr} \pi(f)$ of Corollary 10.7 is taken. Here $f = f^{v,\infty} f_\infty f_{vn}$ is determined by U, π_∞, n as in Corollary 10.7. We rewrite this identity of Corollary 10.7 as follows: for every sufficiently large n, depending on U and ρ (or π'_∞), we have

$$\sum_j c_j u_j^n = \sum_{i, \pi_v} d_i [q_v^{(r-1)/2} z_i(\pi_v)]^n.$$

The two sums are finite. On the right the sum ranges over all components π_v at the fixed place v of the automorphic $G(\mathbb{A})$-modules π with a nonzero U-fixed vector and component π'_∞ at ∞ (these are the π which appear in the identity of Corollary 10.7). The coefficients d_i are positive integers. On the left the u_j are the eigenvalues of the action of the Frobenius $\operatorname{Fr}_v \times 1$ on the $\overline{\mathbb{Q}}_\ell$-vector spaces $H_c^i(\overline{X}, \mathbb{L}(\rho))$ $(0 \leq i \leq 2(r-1))$. Since the coefficients d_i are all positive we conclude the following:

Lemma 10.9. *For every Hecke eigenvalue $z_i(\pi_v)$ of π at v there exists a Frobenius eigenvalue u_j, such that $q_v^{(r-1)/2} z_i(\pi_v) = u_j$.*

Proof. The sums of the last displayed formula are finite (since $H_c^i(\overline{X}, \mathbb{L}(\rho))$ is finite dimensional and there exist only finitely many cuspidal $G(\mathbb{A})$-modules with fixed ramification at all places), and this formula holds for all sufficiently large n. Hence the lemma follows by linear independence of characters. □

To complete the proof of the theorem we need two additional facts.

Lemma 10.10. *Each eigenvalue u_j of the action of the (geometric) Frobenius $\mathrm{Fr}_v \times 1$ on the space $H_c^i(\overline{X}, \mathbb{L}(\rho))$ $(0 \leq i \leq 2(r-1))$ is algebraic, and each complex absolute value of u_j is of the form $q_v^{c/2}$, where c is an integer.*

Proof. This follows from Deligne's theorem [De3] on the integrality of the action of the Frobenius on the cohomology. □

Lemma 10.11. *The complex absolute value $|z_i(\pi_v)|$ of each Hecke eigenvalue $z_i(\pi_v)$ of an unramified component π_v of a cuspidal $G(\mathbb{A})$-module π with a unitary central character satisfies $q_v^{-1/2} < |z_i(\pi_v)| < q_v^{1/2}$.*

Proof. Since π is cuspidal, each of its local components is nondegenerate. By virtue of [Z], (9.7), each such component is equal to a representation induced from the product of a square-integrable representation of a Levi subgroup and an unramified character. If the component π_v is unramified then it is equivalent to an irreducible $I((z_i))$. Since the central character of the cuspidal π is unitary, π is unitary, and each of its components is unitary. By virtue of [B] an irreducible $I((z_i))$ is unitary if and only if for each i there exists a j with $\overline{z}_i^{-1} = z_j$ and $q_v^{-1/2} < |z_i| < q_v^{1/2}$ for every i. This is the required assertion of the lemma. □

The theorem is now an immediate consequence of Lemma 10.9 which compares the integrality result of Lemma 10.10 with the estimate of Lemma 10.11. □

Remark 10.4. Note that the proof of Theorem 10.8 does not show that each Frobenius eigenvalue u_j is related to a Hecke eigenvalue as in Lemma 10.9, since the first sum in Corollary 10.7 is alternating, and an eigenvalue of $\mathrm{Fr}_v \times 1$ on $H_c^i(\overline{X}, \mathbb{L}(\rho))$ may cancel an eigenvalue on $H_c^j(\overline{X}, \mathbb{L}(\rho))$ if $i+j$ is odd.

Remark 10.5 (Field of Definition). As in the paragraph following Prop. 10.4, denote by $L_0(G)$ the space of cusp forms on $G(\mathbb{A})$ which transform under the center Z_∞ of $G(F_\infty)$ according to a character ω. It is well known (see [BJ]) that for each compact open subgroup K of $G(\mathbb{A})$ there exists a compact subset K' of $G(\mathbb{A})$ such that each K-invariant cusp form φ in $L_0(G)$ is supported on $Z_\infty \cdot G(F) \cdot K'$. In particular, the space $L_0(G)_K$ of K-invariant functions φ in $L_0(G)$ is finite dimensional. Denote by $\mathbb{Q}(\omega)$ the field generated by the values of the character ω. It is clear that $L_0(G)$ and $L_0(G)_K$ are defined over $\mathbb{Q}(\omega)$. Let V denote the finite set of places v of F such that $U_v = G(R_v)$ is not contained in K. For every v outside V, the Hecke operators $r(f_v)$ (f_v in $\mathbb{H}_v$) are defined over $\mathbb{Q}(\omega)$, and they commute with each other. Hence they are simultaneously diagonalizable, and their eigenvalues generate a finite extension of $\mathbb{Q}(\omega)$. The eigenspaces are defined over the fields generated by the eigenvalues. Let $\pi = \otimes \pi_v$ be an irreducible $G(\mathbb{A})$-module unramified outside V. Denote by $\mathbb{Q}(\pi_v)$ the field generated by the Hecke eigenvalues of π_v, for v outside V. We conclude that if π is a cuspidal $G(\mathbb{A})$-module, then the compositum $\mathbb{Q}(\pi)$ of $\mathbb{Q}(\pi_v)$ (v outside V), which we call the *field of definition* of π, is a finite extension of $\mathbb{Q}(\omega)$. It is equal to the compositum of $\mathbb{Q}(\pi_v)$ for all v outside V', where V' is any finite set containing V.

Let ρ be an irreducible $D_\infty^\times$-module with finite image, which corresponds to a cuspidal G_∞-module $\pi_\infty(\rho)$. Let $\mathbb{L}(\rho)$ be the smooth $\overline{\mathbb{Q}}_\ell$-adic sheaf on the geometric generic fiber $\overline{M}_{r,I} = M_{r,I} \times_A \overline{F}$ of the moduli scheme $M_{r,I}$ associated with ρ (see Sect. 6.1.4). Let H_ρ^* be the virtual $\mathrm{Gal}(\overline{F}/F) \times \mathbb{H}_I$-module $\sum_i (-1)^i H_c^i(\overline{M}_{r,I}, \mathbb{L}(\rho))$ associated with ρ (see Sect. 6.1.8).

Corollary 10.12. *Suppose that σ is an irreducible ℓ-adic representation of $\mathrm{Gal}(\overline{F}/F)$ which occurs in the virtual module H_ρ^*. Then there exists a finite extension $\mathbb{Q}(\sigma)$ of $\mathbb{Q}$ such that the eigenvalues of $\sigma_v(\mathrm{Fr}_v)$ lie in $\mathbb{Q}(\sigma)$ for all v where σ is unramified.*

Proof. The proof of Theorem 10.8 implies that for almost all v, the (Frobenius) eigenvalues of $\sigma_v(\mathrm{Fr}_v)$ lie in the field $\mathbb{Q}(\pi)$ of definition of a *cuspidal* $G(\mathbb{A})$-module π whose component π_∞ is the cuspidal G_∞-module $\pi_\infty(\rho)$, and π has a nonzero vector fixed by the action of the congruence subgroup U_I. Since there are only finitely many such π, we may take $\mathbb{Q}(\sigma)$ to be the field generated by these finitely many number fields $\mathbb{Q}(\pi)$. $\qquad\square$

11. Existence Theorem

In the proof of Theorem 10.8 we use the Grothendieck fixed point formula of Theorem 6.6, which applies to the cohomology $H_c^i(\overline{X}_v, \mathbb{L}(\rho))$ of the geometric fiber $\overline{X}_v = X_v \otimes_{\mathbb{F}_v} \overline{\mathbb{F}}_v$ of the special fiber $X_v = M_{r,I} \otimes_A \mathbb{F}_v$ (of the moduli scheme $M_{r,I}$), which is a separated scheme of finite type over $\mathbb{F}_v$. This formula applies only to powers of the (geometric) Frobenius endomorphism $\mathrm{Fr}_v \times 1$. Hence the conclusion of Theorem 10.8 concerns only the (Hecke) eigenvalues of the action of the Hecke algebra $\mathbb{H}_v$ of U_v-biinvariant functions on G_v, on this cohomology; as usual we put U_v for $\mathrm{GL}(r, A_v)$.

Our next aim is to study the irreducible constituents which occur (with nonzero multiplicities) in the virtual $\mathrm{Gal}(\overline{F}/F) \times \mathbb{H}_I$-module

$$H = H_\rho^* = \sum_{i=0}^{2(r-1)} (-1)^i H_c^i(\overline{X}, \mathbb{L}(\rho)),$$

where $\overline{X}$ is the geometric generic fiber $M_{r,I} \otimes_A \overline{F}$ of the moduli scheme $M_{r,I}$. Our subsequent results depend on Deligne's conjecture (Theorem 6.8), proven by Fujiwara and Varshavsky. These results assert that each $\mathbb{H}_I$-module which appears in H is automorphic in a sense shortly to be explained, and every automorphic $G(\mathbb{A})$-module with a nonzero U_I-fixed vector and component $\pi_\infty(\rho)$ at ∞ occurs in the space H. Moreover, if $\widetilde{\sigma} \otimes \widetilde{\pi}_f$ is an irreducible constituent of H as a (virtual) $\mathrm{Gal}(\overline{F}/F) \times \mathbb{H}_I$-module, then it occurs with multiplicity one, and the tensor product of $\widetilde{\pi}_f$ with $\nu^{-(r-1)/2}$ (where $\nu(x) = |x|$ for x in $\mathbb{A}^\times$) corresponds to the $\mathrm{Gal}(\overline{F}/F)$-module $\widetilde{\sigma}$, in a sense again to be explained shortly.

The proof depends on the usage of Hecke correspondences in the fixed-point formula, to separate the $\mathbb{H}_I$-modules which appear in H. The scheme $\overline{X}$ is smooth, but not proper. Had $\overline{X}$ been smooth and proper, the Lefschetz fixed-point formula of Theorem 6.7 would apply with any Hecke correspondence. However, this is not the case, and Deligne's conjecture (Theorem 6.8) asserts that although $\overline{X}_v$ is not proper, the fixed-point formula would hold with an arbitrary Hecke correspondence, provided that it is multiplied by a sufficiently high power of the Frobenius. This theorem will be used in conjunction with the trace formula (Prop. 10.6) where the test function $f^{v,\infty}$ outside v and ∞ is arbitrary, and the spherical component f_v at v depends on a parameter m which is sufficiently large with respect to $f^{v,\infty}$ and f_∞. Further, we use [in Lemma 11.4(1)] the congruence relations of Theorem 6.10.

To state the results which depend on Deligne's conjecture (Theorem 6.8), we introduce (as in the first paragraph of Sect. 6.2.2) the following notations. Let π_f be an irreducible $G(\mathbb{A}_f)$-module and $U = U_I$ an open compact congruence subgroup of $G(\mathbb{A}_f)$, defined by a nonzero proper ideal I of the ring A. We denote by π_f^I the (finite dimensional) vector space of U-fixed vectors in π_f. It is naturally an $\mathbb{H}_I$-module. The map $\pi_f \mapsto \pi_f^I$ is a bijection from the set of

Y.Z. Flicker, *Drinfeld Moduli Schemes and Automorphic Forms: The Theory of Elliptic Modules with Applications*, SpringerBriefs in Mathematics, DOI 10.1007/978-1-4614-5888-3_11, © Yuval Z. Flicker 2013

irreducible $G(\mathbb{A}_f)$-modules which have a nonzero U_I-fixed vector, to the set of irreducible $\mathbb{H}_I$-modules in which the unit element of $\mathbb{H}_I$ acts as the identity.

Remark 11.1. This bijection extends to an equivalence of categories, where "irreducible" is replaced by "algebraic."

Definition 11.1. An $\mathbb{H}_I$-module π_f^I is called *cuspidal* if there exists a G_∞-module π_∞ whose product $\pi_f \otimes \pi_\infty$ with the $G(\mathbb{A}_f)$-module π_f which corresponds to π_f^I is a cuspidal $G(\mathbb{A})$-module.

To relate cuspidal $G(\mathbb{A})$-modules with irreducible Galois representations, we need additional definitions.

Definition 11.2. A continuous representation $\sigma : \mathrm{Gal}(\overline{F}/F) \to \mathrm{GL}(r, \overline{\mathbb{Q}}_\ell)$ of the Galois group of F of dimension r is called *constructible* if for almost every place v of F the restriction σ_v of σ to the decomposition group $\mathrm{Gal}(\overline{F}_v/F_v)$ at v is unramified, namely trivial on the inertia subgroup.

If σ_v is unramified then it factorizes through the Galois group $\mathrm{Gal}(\overline{\mathbb{F}}_v/\mathbb{F}_v)$ of the (finite) residue field $\mathbb{F}_v$ of F_v. The group $\mathrm{Gal}(\overline{\mathbb{F}}_v/\mathbb{F}_v)$ is generated by the Frobenius substitution $\varphi : x \mapsto x^{q_v}$ and also by its inverse, the (geometric) Frobenius automorphism $\mathrm{Fr}_v : x \mapsto x^{1/q_v}$ (q_v is the cardinality of $\mathbb{F}_v$). Note that the field $\overline{\mathbb{F}}_v$ is perfect. Suppose that σ_v is semisimple. Then the isomorphism class of σ_v is determined by the eigenvalues (or the conjugacy class) of the matrix $\sigma_v(\mathrm{Fr}_v)$ in $\mathrm{GL}(r, \overline{\mathbb{Q}}_\ell)$, or the characteristic polynomial $P(t; \sigma_v) = \det[t - \sigma_v(\mathrm{Fr}_v)]$ in t.

For any irreducible unramified G_v-module π_v with Hecke eigenvalues $z_i(\pi_v)$ ($1 \le i \le r$), we write $P(t; \pi_v)$ for the product $\prod_{i=1}^r (t - z_i(\pi_v))$. In this chapter we consider only representations over $\overline{\mathbb{Q}}_\ell$, with $\ell \neq p$.

Definition 11.3. A continuous ℓ-adic r-dimensional representation σ of $\mathrm{Gal}(\overline{F}/F)$ and an admissible irreducible $G(\mathbb{A})$-module $\pi = \otimes_v \pi_v$ correspond if σ is constructible and $P(t; \sigma_v) = P(t; \pi_v)$ for almost all v.

As usual we put

$$\overline{X} = M_{r,I} \times_A \overline{F}, \quad \overline{X}_v = M_{r,I} \times_A \overline{F}_v, \quad H_c^i = H_c^i(\overline{X}, \mathbb{L}(\rho)).$$

Theorem 11.1. *Let ρ be an irreducible representation of $D_\infty^\times$ with finite image which corresponds to a cuspidal G_∞-module $\pi_\infty(\rho)$. Let I be a nonzero ideal of A which is contained in at least two maximal ideals and $U = U_I$ the corresponding congruence subgroup of $G(\mathbb{A}_f)$. Put $H^+ = \oplus_i H_c^i$ ($r - 1 - i$ even) and $H^- = \oplus_i H_c^i$ ($r - 1 - i$ odd). Denote by $m_+ = m_+(\widetilde{\sigma} \otimes \widetilde{\pi}_f^I)$ (resp. $m_- = m_-(\widetilde{\sigma} \otimes \widetilde{\pi}_f^I)$) the multiplicity of an irreducible constituent $\widetilde{\sigma} \otimes \widetilde{\pi}_f^I$ of H^+ (resp. H^-) as a $\mathrm{Gal}(\overline{F}/F) \times \mathbb{H}_I$-module, and put $m = m_+ - m_-$. Then (1) $m(\widetilde{\sigma} \otimes \widetilde{\pi}_f^I)$ is equal to zero or one. (2) If $m(\widetilde{\sigma} \otimes \widetilde{\pi}_f^I)$ is one then the corresponding $G(\mathbb{A})$-module $\widetilde{\pi} = \widetilde{\pi}_f \otimes \pi_\infty(\rho)$ is cuspidal (in particular automorphic), and the dimension of $\widetilde{\sigma}$ is r. Put $\nu(x) = |x|$ (x in $\mathbb{A}^\times$). (3) If $m(\widetilde{\sigma} \otimes \widetilde{\pi}_f^I)$ is one then $\widetilde{\pi} \otimes \nu^{-(r-1)/2}$ corresponds to $\widetilde{\sigma}$. (4) For every cuspidal $G(\mathbb{A})$-module $\pi = \pi_f \otimes \pi_\infty$ with $\pi_\infty = \pi_\infty(\rho)$ and $\pi_f^I \neq \{0\}$ there exists a $\mathrm{Gal}(\overline{F}/F) \times \mathbb{H}_I$-module $\widetilde{\sigma} \otimes \widetilde{\pi}_f^I$ with $m(\widetilde{\sigma} \otimes \widetilde{\pi}_f^I) = 1$ such that $\pi_f^I \simeq \widetilde{\pi}_f^I$.*

Combining (3) and (4) we deduce the following corollary, which is used in a crucial way to establish in Chap. 12 the higher reciprocity law.

Corollary 11.2. *For every cuspidal $G(\mathbb{A})$-module $\pi = \otimes_v \pi_v$ such that π_∞ is cuspidal and its central character is of finite order, there exists a continuous irreducible r-dimensional ℓ-adic constructible representation σ of $\mathrm{Gal}(\overline{F}/F)$ which corresponds to π.*

Proof. This follows from Theorem 11.1 on taking ρ which corresponds to π_∞ and a sufficiently small congruence subgroup $U = U_I$ of $G(\mathbb{A}_f)$ such that $\pi_f^I \neq \{0\}$. $\square$

Proof of Theorem. The cohomology space $H_c^i(\overline{X}_v, \mathbb{L}(\rho))$ is a $\mathrm{Gal}(\overline{\mathbb{F}}_v/\mathbb{F}_v) \times \mathbb{H}_I$-module for every maximal ideal v of A not containing I. As noted in (6.1.7), the constructibility of the sheaf $\mathbb{L}(\rho)$ implies that for almost all places $v \neq \infty$ (in F) which do not divide I, the restriction of $H_c^i(\overline{X}, \mathbb{L}(\rho))$ to the decomposition group $\mathrm{Gal}(\overline{F}_v/F_v)$ at v is trivial on the inertia subgroup and is isomorphic, as a $\mathrm{Gal}(\overline{\mathbb{F}}_v/\mathbb{F}_v) \times \mathbb{H}_I$-module, to $H_c^i(\overline{X}_v, \mathbb{L}(\rho))$. We prove the theorem on applying Deligne's conjecture to the $H_c^i(\overline{X}_v, \mathbb{L}(\rho))$ for the set of these v.

For any i ($0 \leq i \leq 2(r-1)$) denote by $\widetilde{\sigma}_i \otimes \widetilde{\pi}_{i,f}^I$ the irreducible constituents, repeated according to their multiplicities, of the $\mathrm{Gal}(\overline{F}/F) \times \mathbb{H}_I$-module $H_c^i(\overline{X}, \mathbb{L}(\rho))$. The spaces $H_c^i(\overline{X}, \mathbb{L}(\rho))$ and $H_c^i(\overline{X}_v, \mathbb{L}(\rho))$ are isomorphic as $\mathrm{Gal}(\overline{F}_v/F_v) \times \mathbb{H}_I$-modules for almost all v, and the restriction $\widetilde{\sigma}_{i,v}$ of $\widetilde{\sigma}_i$ to the decomposition group $\mathrm{Gal}(\overline{F}_v/F_v)$ factorizes through the quotient $\mathrm{Gal}(\overline{\mathbb{F}}_v/\mathbb{F}_v)$.

Let f^∞ be an element of $\mathbb{H}_I$. It is a compactly supported U-biinvariant $\overline{\mathbb{Q}}_\ell$-valued function on $G(\mathbb{A}_f)$. Then f^∞ defines a correspondence on $\overline{X}_v$ and an automorphism of $H_c^i(\overline{X}_v, \mathbb{L}(\rho))$ as in Sect. 6.1.8, which will also be denoted here by f^∞. For every such f^∞ and integers m and $i \geq 0$, we have

$$(11.1) \qquad \mathrm{tr}[(\mathrm{Fr}_v^m \times 1) \cdot f^\infty; H_c^i(\overline{X}_v, \mathbb{L}(\rho))] = \sum \mathrm{tr}\, \widetilde{\pi}_{i,f}(f^\infty)\, \mathrm{tr}\, \widetilde{\sigma}_{i,v}(\mathrm{Fr}_v^m);$$

on the right the sum ranges over the irreducible constituents $\widetilde{\sigma}_i \otimes \widetilde{\pi}_{i,f}$ of $H_c^i(\overline{X}, \mathbb{L}(\rho))$ as a $\mathrm{Gal}(\overline{F}/F) \times \mathbb{H}_I$-module.

Suppose in addition that $f^\infty = f^{v,\infty} f_v^0$, where $f^{v,\infty}$ is a function on $G(\mathbb{A}_f^v)$ (where $\mathbb{A}_f^v$ is the ring of adèles without components at v and ∞) and f_v^0 is the unit element of the Hecke algebra $\mathbb{H}_v$ with respect to $U_v = G(A_v)$.

Proposition 10.4 implies that for every f^∞ as above there exists an integer m_0 such that for every integer $m \geq m_0$, the alternating sum

$$(11.2) \qquad \sum_{i=0}^{2(r-1)} (-1)^{r-1-i}\, \mathrm{tr}[(\mathrm{Fr}_v^m \times 1) \cdot f^\infty; H_c^i(\overline{X}_v, \mathbb{L}(\rho))]$$

of the left sides of Eq. (11.1) is equal to the geometric side

$$(11.3) \qquad \sum_{\{\gamma\}} |Z_\gamma(G(\mathbb{A}))/Z_\gamma(G(F))Z_\infty| \cdot \Phi(\gamma, f_\infty \cdot f^\infty \cdot f_{m,v})$$

of the trace formula. Here, as in Prop. 10.3, f_∞ is a normalized matrix coefficient of the cuspidal G_∞-module $\pi_\infty(\rho)$ which corresponds to the $D_\infty^\times$-module ρ. As usual, $f_{m,v}$ denotes the spherical function on G_v of Definition 9.1. The sum in Eq. (11.3) ranges over the set of conjugacy classes of the elliptic γ in $G = G(F)$.

The trace formula of Prop. 10.6 asserts that Eq. (11.3) is equal to

$$(11.4) \qquad \sum_\pi \operatorname{tr} \pi_v(f_{m,v}) \operatorname{tr} \pi^\infty(f^\infty) \operatorname{tr} \pi_\infty(f_\infty)$$

for all $m \geq m_0 = m_0(f^\infty f_\infty)$. The sum ranges over all cuspidal $G(\mathbb{A})$-modules $\pi = \otimes_w \pi_w$. Note that (i) if the component π_∞ satisfies $\operatorname{tr} \pi_\infty(f_\infty) \neq 0$, then π_∞ is the cuspidal $\pi_\infty(\rho)$ and $\operatorname{tr} \pi_\infty(f_\infty) = 1$; (ii) if $\operatorname{tr} \pi^\infty(f^\infty) \neq 0$, then $\pi^\infty = \otimes_w \pi_w$ ($w \neq \infty$) has a nonzero U-fixed vector. Consequently, the sum of Eq. (11.4) is finite since there are only finitely many cuspidal $G(\mathbb{A})$-modules π with a nonzero U-fixed vector and the component $\pi_\infty(\rho)$ at ∞.

Denote by $u_{j,v}(\widetilde{\sigma}_i)$ the eigenvalues of the matrix $\widetilde{\sigma}_{i,v}(\mathrm{Fr}_v)$. By virtue of Eq. (11.1) the alternating sum of Eq. (11.2) is equal to

$$(11.5) \qquad \sum_{i=0}^{2(r-1)} (-1)^{r-1-i} \sum_{\widetilde{\sigma}_i \otimes \widetilde{\pi}_{i,f}} \operatorname{tr} \widetilde{\pi}_{i,f}(f^\infty) \left[\sum_j u_{j,v}(\widetilde{\sigma}_i)^m \right].$$

Denoting as usual the Hecke eigenvalues of π_v by $z_{j,v}(\pi_v)$ ($1 \leq j \leq r$), we rewrite Eq. (11.4) in the form

$$(11.6) \qquad \sum_\pi \operatorname{tr} \pi^\infty(f^\infty) \left[\sum_{j=1}^r (q_v^{(r-1)/2} z_{j,v}(\pi))^m \right].$$

Denote by $\widetilde{\pi}^\infty$ a $G(\mathbb{A}_f)$-module (up to equivalence) which contributes to Eq. (11.5). Let $\widetilde{\sigma}^+ = \widetilde{\sigma}^+(\widetilde{\pi}^\infty)$ be the sum of the $\widetilde{\sigma}_i$ over the $\widetilde{\pi}^\infty \otimes \widetilde{\sigma}_i$ which occur in H^+ and define $\widetilde{\sigma}^- = \widetilde{\sigma}^-(\widetilde{\pi}^\infty)$ similarly, using H^-. Denote the corresponding eigenvalues by $u_{j,v}(\widetilde{\sigma}^+(\widetilde{\pi}^\infty))$ and $u_{j,v}(\widetilde{\sigma}^-(\widetilde{\pi}^\infty))$. Then Eq. (11.5) can be rewritten in the form

$$(11.7) \qquad \sum_{\widetilde{\pi}^\infty} \operatorname{tr} \widetilde{\pi}^\infty(f^\infty) \left(\sum_j u_{j,v}(\widetilde{\sigma}^+(\widetilde{\pi}^\infty))^m - \sum_j u_{j,v}(\widetilde{\sigma}^-(\widetilde{\pi}^\infty))^m \right).$$

Fix an element $\boldsymbol{a}$ in $\overline{\mathbb{Q}}_\ell^{\times r}/S_r$. Denote by $(a_i; 1 \leq i \leq r)$ its multiset (set with repetitions) of components. Since the sums Eqs. (11.6) and (11.7) are finite and equal for all $m \geq m_0(f^\infty)$, we conclude that

$$(11.8) \qquad \sum_\pi \operatorname{tr} \pi^\infty(f^\infty) = \sum_{\widetilde{\pi}^\infty} \operatorname{tr} \widetilde{\pi}^\infty(f^\infty);$$

on the left π ranges over the set of π in Eq. (11.6) such that the image of $(z_{j,v}(\pi))$ in $\mathbb{Q}_\ell^{\times r}/S_r$ is $q_v^{(1-r)/2}\boldsymbol{a}$. On the right $\widetilde{\pi}^\infty$ ranges over a set of $\widetilde{\pi}^\infty$ in Eq. (11.7) with $u_{j,v}(\widetilde{\sigma}^*(\widetilde{\pi}^\infty))$ in $\{a_j\}$. In Eq. (11.8) the function f^∞ is an arbitrary element of $\mathbb{H}_I$ of the form $f^{v,\infty} f_v^0$. Since the sums of Eq. (11.8) are finite we conclude that for every $\tilde{\pi}^\infty = \tilde{\pi}^{v,\infty} \otimes \tilde{\pi}_v$ there is a necessarily unique automorphic $\pi = \pi_\infty \otimes \pi^{v,\infty} \otimes \pi_v$ with $\tilde{\pi}^{v,\infty} \simeq \pi^{v,\infty}$. Since we may vary v we conclude that $\tilde{\pi}^\infty \simeq \pi^\infty$ and that $\tilde{\pi} = \tilde{\pi}^\infty \otimes \pi_\infty(\rho)$ is automorphic. Moreover, for almost all v and for all m, we have

$$(11.9) \qquad \sum_j u_{j,v}(\tilde{\sigma}^+(\tilde{\pi}))^m - \sum_j u_{j,v}(\tilde{\sigma}^-(\tilde{\pi}))^m = \sum_{j=1}^r (q_v^{(r-1)/2} z_{j,v}(\pi))^m.$$

In summary, as a virtual $\mathrm{Gal}(\overline{F}/F) \times \mathbb{H}_I$-module, $H^+ - H^-$ is the sum over all cuspidal $G(\mathbb{A})$-modules $\pi = \pi_f \otimes \pi_\infty(\rho)$ with $\pi_f^I \neq \{0\}$ of $\sigma \otimes \pi_f^I$, where $\sigma = \sigma(\pi)$ is a virtual representation of $\mathrm{Gal}(\overline{F}/F)$. Thus $\sigma = \sum_{j \geq 0} m_j \sigma_j$, where the σ_j are irreducible and the (finitely many) m_j are nonzero integers. Put σ^* for $\sigma \otimes \nu^{(r-1)/2}$, where ν is the character of $\mathrm{Gal}(\overline{F}/F)$ which corresponds to $\nu(x) = |x|$ (x in $\mathbb{A}^\times$) by class field theory. Put $P(t; \sigma(\pi)_v^*) = \prod_j P(t; \sigma_j(\pi)_v^*)^{m_j}$. We conclude the following intermediate result.

Lemma 11.3. *There is a finite set V of places of F, including ∞ and the divisors of I, such that the $\sigma_j = \sigma_j(\pi)$ are unramified at each v outside V and $P(t; \sigma(\pi)_v^*) = P(t; \pi_v)$ for all v outside V and for all π.*

To complete the proof of the theorem we need to show that each σ is irreducible, namely that $m_0 = 1$ and $m_j = 0$ for $j > 0$.

Lemma 11.4. *(1) Let $\tilde{\sigma} \otimes \tilde{\pi}_f^I$ be an irreducible constituent of the $\mathrm{Gal}(\overline{F}/F) \times \mathbb{H}_I$-module H_c^i for some i ($0 \leq i \leq 2(r-1)$). Then for almost all v, each Frobenius eigenvalue $u_{j,v}(\tilde{\sigma}^*)$ of $\tilde{\sigma}^*$ is equal to some Hecke eigenvalue $z_{k,v}(\tilde{\pi}_f)$ ($k = k(j)$) of $\tilde{\pi}_f$. (2) The complex absolute value of each conjugate of the algebraic number $z_{k,v}(\tilde{\pi}_f)$ is one.*

Proof. (1) is Theorem 6.10 (congruence relations); (2) is Theorem 10.8 (Hecke purity). □

Combining the two parts of Lemma 11.4 we obtain the following Frobenius purity result.

Lemma 11.5. *For each v outside V and σ_j as in Lemma 11.3, each conjugate of each Frobenius eigenvalue $u_{i,v}(\sigma_j^*)$ of σ_j^* has complex absolute value one.*

Consequently the $\overline{\mathbb{Q}}_\ell$-adic sheaf $S(\sigma_j^*)$ on the curve C ($F = \mathbb{F}_q(C)$), which is associated (see [De2], Section 10) to the constructible irreducible $\overline{\mathbb{Q}}_\ell$-adic representation σ_j^* of $\mathrm{Gal}(\overline{F}/F)$, is (smooth on an open dense subscheme $C(\sigma_j^*)$ of C and) pure of weight zero in the terminology of Deligne [De3] (see also [SGA4 1/2]; Sommes trig., pp. 177/8).

To prove the irreducibility of σ we use basic properties of L-functions. For any virtual representation σ of $\mathrm{Gal}(\overline{F}/F)$ which is unramified outside V put

$$L(s, \sigma_v) = \prod_i (1 - q_v^{-s} u_i(\sigma_v))^{-1} \qquad \text{and} \qquad L(s, V, \sigma) = \prod_{v \notin V} L(s, \sigma_v).$$

The absolute convergence of the product $L(s, V, \sigma_j^*)$ in the right half plane $\mathrm{Re}(s) > 1$ follows from Lemma 11.5. Then

$$L\left(s, V, \left(\sum_i m_i \sigma_i\right) \otimes \left(\sum_j m_j' \sigma_j'\right)\right) = \prod_{i,j} L(s, V, \sigma_i \otimes \sigma_j')^{m_i m_j'}.$$

Now Grothendieck proved that the product $L(s, V, \sigma)$ is a rational function in q^s on identifying it (see [De2], Section 10) with an L-function $L(s, S(\sigma, V)/X)$ of a smooth $\overline{\mathbb{Q}}_\ell$-adic sheaf $S(\sigma, V)$ over the scheme $X = \mathrm{Spec}\, A - V$. If S is a smooth sheaf on a scheme X_d of dimension d over $\mathbb{F}_q$ then the L-function is a product

$$L(s, S/X_d) = \prod_{0 \leq j \leq 2d} P_j(s, S/X_d)^{(-1)^{j+1}},$$

where $P_j(s, S/X_d)$ are polynomials associated with $H_c^j(X_d \times_{\mathbb{F}_q} \overline{\mathbb{F}}_q, S)$. The results of Deligne [De3] assert that if $S = S_i$ is pure of weight i then the zeroes of $P_j(s, S_i/X_d)$ occur for half-integral $\mathrm{Re}\, s \leq (i + j)/2$.

In our case the sheaf $S_0 = S(\sigma_j^*, V)$ associated with any σ_j^* as in Lemma 11.5 is pure of weight zero, over a curve $X_1 = C(\sigma_j^*)$. Thus $d = 1$. Hence we conclude

Lemma 11.6. *For every σ_j as in Lemma 11.3, the rational function $L(s, V, \sigma_j^*)$ has no pole in $\mathrm{Re}\, s > 1$ and no zero in $\mathrm{Re}\, s > \frac{1}{2}$.*

On the other hand we introduce the functions

$$L(s, \pi_v \otimes \pi_v') = \prod_{i,j} (1 - q_v^{-s} z_i(\pi_v) z_j(\pi_v'))^{-1}, \quad L(s, \pi \otimes \pi') = \prod_{v \notin V} L(s, \pi_v \otimes \pi_v'),$$

for any cuspidal $G(\mathbb{A})$-modules π, π' which are unramified outside V. It is easy to see that the infinite product converges in some right half plane ($\mathrm{Re}\, s > c(> 1)$). By virtue of [JS], Prop. 10.6, we further have

Lemma 11.7. *The product $L(s, V, \pi \otimes \pi')$ is a rational function in q^s which is regular in $\mathrm{Re}\, s > 1$. It has a pole at $s = 1$, which is necessarily simple, if and only if π' is the contragredient $\pi^\vee$ of π.*

Note also that if σ is the trivial representation of $\mathrm{Gal}(\overline{F}/F)$ then $L(s, V, \sigma)$ has a pole at $s = 1$. If σ is irreducible and its contragredient is denoted by $\sigma^\vee$, then $\sigma \otimes \sigma^\vee$ contains a copy of the trivial representation. Hence $L(s, V, \sigma \otimes \sigma^\vee)$ has a pole at $s = 1$.

Now suppose that $\sigma \otimes \pi_f^I$ occurs in Lemma 11.3, where $\sigma = \sum_j m_j \sigma_j$, the σ_j are irreducible and the m_j are integral. Then

$$L(s, V, \pi \otimes \pi^\vee) = L\left(s, V, \left(\sum_j m_j \sigma_j^*\right) \otimes \left(\sum_j m_j \sigma_j^{\vee *}\right)\right) = \prod_{i,j} L(s, V, \sigma_i^* \otimes \sigma_j^{\vee *})^{m_i m_j}.$$

The order of the pole of the left side at $s = 1$ is one, while the order of pole of the right side at $s = 1$ is at least $\sum_i m_i^2$, since $L(s, V, \sigma_i^* \otimes \sigma_j^{\vee *})$ does not vanish at $s = 1$ for all i, j, by Lemma 11.6. We conclude that $m_i = 0$ for $i > 0$ and $m_0^2 = 1$ (on rearranging indices). But then it is clear that $m_0 = 1$, e.g., from Eq. (11.9). This completes the reduction of the theorem to Deligne's conjecture (Theorem 6.8).

12. Representations of a Weil Group

Let $F = \mathbb{F}_q(C)$ be the field of functions on a smooth projective absolutely irreducible curve C over $\mathbb{F}_q$, $\mathbb{A}$ its ring of adèles, $\overline{F}$ a separable algebraic closure of F, $G = \mathrm{GL}(r)$, and ∞ a fixed place of F, as in Chap. 2. This chapter concerns the higher reciprocity law, which parametrizes the cuspidal $G(\mathbb{A})$-modules whose component at ∞ is cuspidal, by irreducible continuous constructible r-dimensional ℓ-adic ($\ell \neq p$) representations of the Weil group $W(\overline{F}/F)$, or irreducible rank r smooth ℓ-adic sheaves on $\mathrm{Spec}\, F$ which extend to smooth sheaves on an open subscheme of the smooth projective curve whose function field is F, whose restriction to the local Weil group $W(\overline{F}_\infty/F_\infty)$ at ∞ is irreducible. This law is reduced to Theorem 11.1, which depends on Deligne's conjecture (Theorem 6.8). This reduction uses the Converse Theorem 13.1 and properties of ε-factors attached to Galois representations due to Deligne [De2] and Laumon [Lm1]. We explain the result twice. A preliminary exposition is in the classical language of representations of the Weil group. Then in the equivalent language of smooth ℓ-adic sheaves, used, e.g., in [DF]. Note that in this chapter we denote a Galois representation by ρ, as σ is used to denote an element of a Galois group.

12.1. **Weil Groups.** Let F be a local non-Archimedean field with ring R of integers, residue field $\mathbb{F}$, and separable closure $\overline{F}$. Let $\overline{\mathbb{F}}$ denote the residue field of the integral closure $\overline{R}$ of R in $\overline{F}$. Then $\overline{\mathbb{F}}$ is an algebraic closure of the finite field $\mathbb{F}$. The kernel I of the natural epimorphism $\mathrm{Gal}(\overline{F}/F) \to \mathrm{Gal}(\overline{\mathbb{F}}/\mathbb{F})$ is called the *inertia* subgroup of $\mathrm{Gal}(\overline{F}/F)$. The Galois group $\mathrm{Gal}(\overline{\mathbb{F}}/\mathbb{F})$ is isomorphic to the profinite completion $\widehat{\mathbb{Z}} = \lim_n \mathbb{Z}/n$ of $\mathbb{Z}$. It is topologically generated by the arithmetic Frobenius automorphism $\varphi : x \mapsto x^q$ of $\overline{\mathbb{F}}$, where q is the cardinality of $\mathbb{F}$. Since φ is bijective, we can and do introduce also the (geometric) Frobenius morphism $\mathrm{Fr} = \varphi^{-1}$. The *Weil* group $W(\overline{F}/F)$ is the group of g in $\mathrm{Gal}(\overline{F}/F)$ whose image in $\mathrm{Gal}(\overline{\mathbb{F}}/\mathbb{F})$ is an integral power of Fr. Let $W(\overline{\mathbb{F}}/\mathbb{F})$ denote the group $\langle \mathrm{Fr}^n; n \text{ in } \mathbb{Z}\rangle \simeq \mathbb{Z}$. Then there is an exact sequence

$$1 \to I \to W(\overline{F}/F) \to W(\overline{\mathbb{F}}/\mathbb{F}) \simeq \mathbb{Z} \to 0.$$

The Galois group $\mathrm{Gal}(\overline{F}/F)$ is a topological group, in the topology where a system of neighborhoods of the identity is given by $\mathrm{Gal}(\overline{F}/F')$, where F' ranges through the set of finite extensions of F in $\overline{F}$. Then $\mathrm{Gal}(\overline{F}/F) = \lim_{\leftarrow} \mathrm{Gal}(F'/F)$ is a profinite group, hence compact. The Weil group $W(\overline{F}/F)$ is given the topology where a fundamental system of neighborhoods of the identity is the same as in I. The group $\mathrm{Gal}(\overline{F}/F)$ is the profinite completion of $W(\overline{F}/F)$. The subgroup of $W(\overline{F}/F)$ corresponding to the finite extension F' of F is identified with $W(\overline{F}/F')$. Let $\deg : F \twoheadrightarrow \mathbb{Z}$ denote the normalized additive valuation.

Y.Z. Flicker, *Drinfeld Moduli Schemes and Automorphic Forms: The Theory*   113
of Elliptic Modules with Applications, SpringerBriefs in Mathematics,
DOI 10.1007/978-1-4614-5888-3_12, © Yuval Z. Flicker 2013

Local class field theory implies that there is a commutative diagram of topological groups:

$$
\begin{array}{ccccccccc}
1 & \longrightarrow & I & \longrightarrow & W(\overline{F}/F) & \longrightarrow & W(\overline{\mathbb{F}}/\mathbb{F}) & \longrightarrow & 1 \\
 & & \downarrow & & \downarrow & & \downarrow\wr & & \\
1 & \longrightarrow & R^{\times} & \longrightarrow & F^{\times} & \underset{\deg}{\longrightarrow} & \mathbb{Z} & \longrightarrow & 0,
\end{array}
$$

such that the reciprocity homomorphism $W(\overline{F}/F) \twoheadrightarrow F^{\times}$ is surjective, and its kernel consists of the commutator subgroup of $W(\overline{F}/F)$. It is normalized so that the (geometric) Frobenius is mapped to a uniformizer in R. Consequently the quotient $W(\overline{F}/F)_{\mathrm{ab}}$ is isomorphic to $F^{\times}$, and there is a natural bijection between the sets of continuous one-dimensional representations of $W(\overline{F}/F)$ and of $\mathrm{GL}(1, F)$. An *unramified* representation of $W(\overline{F}/F)$ is one which is trivial on the inertia subgroup I. An unramified character of $W(\overline{F}/F)$ corresponds to an unramified character of $F^{\times}$.

Let F be a function field in one variable over $\mathbb{F}_p$ and $\mathbb{F} = \mathbb{F}_q$ its subfield of constants ($=$ algebraic closure of $\mathbb{F}_p$ in F). Let $\overline{F}$ be a separable closure of F, $\overline{\mathbb{F}}$ the algebraic closure of $\mathbb{F}$ in $\overline{F}$, and $\mathrm{Gal}(\overline{F}/F)_0$ the kernel of the restriction homomorphism from $\mathrm{Gal}(\overline{F}/F)$ to $\mathrm{Gal}(\overline{\mathbb{F}}/\mathbb{F})$. The global *Weil group* $W(\overline{F}/F)$ is defined by the diagram

$$
\begin{array}{ccccccccc}
1 & \longrightarrow & \mathrm{Gal}(\overline{F}/F)_0 & \longrightarrow & W(\overline{F}/F) & \longrightarrow & W(\overline{\mathbb{F}}/\mathbb{F}) \simeq \mathbb{Z} & \longrightarrow & 0 \\
 & & \wr\downarrow & & \downarrow & & \downarrow & & \\
1 & \longrightarrow & \mathrm{Gal}(\overline{F}/F)_0 & \longrightarrow & \mathrm{Gal}(\overline{F}/F) & \longrightarrow & \mathrm{Gal}(\overline{\mathbb{F}}/\mathbb{F}) \simeq \widehat{\mathbb{Z}} & \longrightarrow & 0.
\end{array}
$$

Global class field theory yields an isomorphism of $W(\overline{F}/F)_{\mathrm{ab}}$ with the idèle class group $\mathbb{A}^{\times}/F^{\times}$ and in particular a natural isomorphism from the set of continuous one-dimensional representations of $W(\overline{F}/F)$ to the set of automorphic representations of $\mathrm{GL}(1, \mathbb{A})$. Let v be a place of F and $\overline{v}$ a place of $\overline{F}$ over v. The (decomposition) subgroup $D_{\overline{v}}$ of $W(\overline{F}/F)$, consisting of all w with $w\,\overline{v} = \overline{v}$, is isomorphic to the local Weil group $W(\overline{F}_{\overline{v}}/F_v)$, where the completion $\overline{F}_{\overline{v}}$ of $\overline{F}$ at $\overline{v}$ is a separable closure of F_v. The quotient of $W(\overline{F}_{\overline{v}}/F_v)$ by the inertia subgroup $I_{\overline{v}}$ is isomorphic to the subgroup $W(\overline{\mathbb{F}}_{\overline{v}}/\mathbb{F}_v) \simeq \mathbb{Z}$ of $\mathrm{Gal}(\overline{\mathbb{F}}_{\overline{v}}/\mathbb{F}_v) \simeq \widehat{\mathbb{Z}}$, generated by the Frobenius. The quotient $\mathbb{F}_v = R_v/v$ is the residue field of F_v. The local and global Weil groups are related by the diagram

$$
\begin{array}{ccccccccc}
1 & \longrightarrow & I_{\overline{v}} & \longrightarrow & W(\overline{F}_{\overline{v}}/F_v) & \longrightarrow & W(\overline{\mathbb{F}}_{\overline{v}}/\mathbb{F}_v) \simeq \mathbb{Z} & \longrightarrow & 0 \\
 & & \downarrow & & \downarrow & & \downarrow & & \\
1 & \longrightarrow & \mathrm{Gal}(\overline{F}/F)_0 & \longrightarrow & W(\overline{F}/F) & \longrightarrow & W(\overline{\mathbb{F}}/\mathbb{F}) \simeq \mathbb{Z} & \longrightarrow & 0.
\end{array}
$$

The vertical arrow on the right is multiplication by $[\mathbb{F}_v : \mathbb{F}]$. The local and global class field theories are related by the commutative diagram

$$
\begin{array}{ccc}
W(\overline{F}_{\overline{v}}/F_v)_{\mathrm{ab}} & \longrightarrow & W(\overline{F}/F)_{\mathrm{ab}} \\
\wr\downarrow & & \downarrow\wr \\
F_v^{\times} & \longrightarrow & \mathbb{A}^{\times}/F^{\times}.
\end{array}
$$

12.2. ℓ-Adic Representations.

Let $\ell \neq p$ be a rational prime, E_λ a finite field extension of $\mathbb{Q}_\ell$, and V_λ an r-dimensional vector space over E_λ. The topology on the group $\operatorname{Aut} V_\lambda \simeq \operatorname{GL}(r, E_\lambda)$ is induced by that of $\operatorname{End} V_\lambda \simeq E_\lambda^{r^2}$. A λ-*adic representation* of F is a continuous homomorphism $\rho : W(\overline{F}/F) \to \operatorname{Aut} V_\lambda$. The restriction to $W(\overline{F}/F)$ of a λ-adic representation $\widetilde{\rho} : \operatorname{Gal}(\overline{F}/F) \to \operatorname{Aut} V_\lambda$ of the Galois group $\operatorname{Gal}(\overline{F}/F)$ is a λ-adic representation. But not every representation of $W(\overline{F}/F)$ extends to a representation of $\operatorname{Gal}(\overline{F}/F)$. Since $W(\overline{F}/F)$ is topologically finitely generated over the profinite (and consequently compact) group $\operatorname{Gal}(\overline{F}/F)_0$, every continuous ℓ-adic representation $\rho : W(\overline{F}/F) \to \operatorname{GL}(r, \overline{\mathbb{Q}}_\ell)$ factorizes through $\operatorname{GL}(r, E_\lambda)$ for some finite extension E_λ of $\mathbb{Q}_\ell$ in $\overline{\mathbb{Q}}_\ell$. Indeed, this follows from

Lemma 12.1. *For every compact subgroup K of $H = \operatorname{GL}(r, \overline{\mathbb{Z}}_\ell)$ there is a finite extension E_λ of $\mathbb{Q}_\ell$ in $\overline{\mathbb{Q}}_\ell$ such that K lies in $\operatorname{GL}(r, E_\lambda)$.*

Proof. Here $\overline{\mathbb{Z}}_\ell$ is the ring of integers in $\overline{\mathbb{Q}}_\ell$, and by M_r we denote the ring of $r \times r$ matrices. The congruence subgroup $H_i = I + \ell^i M_r(\overline{\mathbb{Z}}_\ell)$ ($i \geq 1$) is an open (normal) subgroup of H. Hence $K_i = K \cap H_i$ is an open subgroup of K, and so the quotient K/K_i is discrete and compact, and finite. If K_i is contained in $\operatorname{GL}(r, E)$ for some finite extension E of $\mathbb{Q}_\ell$, then K lies in $\operatorname{GL}(r, E')$, where E' is generated over E by the coefficients of a set of coset representatives in K for K/K_i; since K/K_i is finite, so is $[E' : \mathbb{Q}_\ell]$. Suppose that K is not contained in $\operatorname{GL}(r, E)$ for any finite E over $\mathbb{Q}_\ell$, then K_i is not contained in $\operatorname{GL}(r, E)$ for any finite $E/\mathbb{Q}_\ell$ and any i. There exists a sequence $n_1 < n_2 < \ldots$ of positive integers and elements g_i in K_{n_i}, satisfying:

1. If $\sigma(g_i) \neq g_i$ then $\sigma(g_i) \not\equiv g_i \bmod \ell^{n_i+1}$, for any σ in $\operatorname{Gal}(\overline{\mathbb{Q}}_\ell/\mathbb{Q}_\ell)$.
2. The field generated over $\mathbb{Q}_\ell$ by the entries in the matrix g_i has degree $\geq i$ over $\mathbb{Q}_\ell$.

Indeed, g_1 can be chosen arbitrarily; once $g_1, \ldots, g_{i-1}$ (and $n_1, \ldots, n_{i-1}$) are chosen, since g_{i-1} has only finitely many conjugates, n_i ($> n_{i-1}$) can be chosen to satisfy (1) and g_i can be chosen in K_{n_i} to satisfy (2).

Put $h = g_1 g_2 \ldots$; the product converges to an element h in K since $n_i \to \infty$. For an automorphism σ of $\overline{\mathbb{Q}}_\ell$ over $\mathbb{Q}_\ell$ which fixes h one has $g_1 g_2 \ldots = \sigma(g_1)\sigma(g_2)\ldots$. Denote by j the least $i \geq 1$ with $\sigma(g_i) \neq g_i$ (it is clear from (2) that there is such $j < \infty$). Then $g_j g_{j+1} \ldots = \sigma(g_j)\sigma(g_{j+1})\ldots$. Hence $\sigma(g_j) \equiv g_j \bmod \ell^{n_j+1}$, but $\sigma(g_j) \neq g_j$; this is a contradiction to (1) which proves the lemma. $\qquad \square$

Definition 12.1. A λ-adic representation ρ of F is called *unramified* at the place v of F if $\rho(I_{\overline{v}}) = \{1\}$ for some, hence for every, place $\overline{v}$ of $\overline{F}$ which extends v.

In this case the restriction of ρ to $D_{\overline{v}} = W(\overline{F}_{\overline{v}}/F_v)$ factorizes through $D_{\overline{v}}/I_{\overline{v}} \simeq \langle \operatorname{Fr}_{\overline{v}} \rangle$, and the image $\rho(\operatorname{Fr}_{\overline{v}})$ of the (geometric) Frobenius is well defined. The conjugacy class of $\rho(\operatorname{Fr}_{\overline{v}})$ in $\operatorname{GL}(r, E_\lambda)$ is independent of the choice of $\overline{v}$; it is denoted by $\rho(\operatorname{Fr}_v)$. For any $\overline{v}$ over v let $V_\lambda^{\rho(I_{\overline{v}})}$ be the space of $\rho(I_{\overline{v}})$-fixed vectors in V_λ. The characteristic polynomial

$$P_{v,\rho}(t) = \det[1 - t \cdot \rho(\mathrm{Fr}_{\overline{v}}) \mid V_\lambda^{\rho(I_{\overline{v}})}]$$

is independent of the choice of $\overline{v}$.

Definition 12.2. A λ-adic representation of F is called *constructible* if it is unramified at almost all v.

It would be interesting to show that every semisimple, in particular irreducible, λ-adic representation of F is constructible. However, this is not yet known, and from now on by a λ-*adic representation* ρ of F we mean one which is constructible, namely one such that ρ_v is unramified for almost all v.

Example 12.1. There does exist a two-dimensional indecomposable reducible nonsemisimple ℓ-adic representation ρ which is ramified at each place of F. To see this, denote the distinct monic irreducible polynomials on the curve C defining $F = \mathbb{F}_q(C)$ by p_0, $p_1,\ldots$. Put x_n for a root of the equation $x^{\ell^n} = p_0 p_1^\ell \cdots p_i^{\ell^i} \cdots p_{n-1}^{\ell^{n-1}}$, and fix a primitive ℓ^n-th root ζ_n of 1 for all n, say with $\zeta_n^\ell = \zeta_{n-1}$. For each σ in $\mathrm{Gal}(\overline{F}/F)$, where $\overline{F}$ is a separable closure of F, define $\chi_\ell(\sigma)$ and $\alpha_\ell(\sigma)$ in $\mathbb{Z}_\ell$ by $\sigma\zeta_n = \zeta_n^{\chi_\ell(\sigma)(\mathrm{mod}\ \ell^n)}$ and $\sigma x_n = \zeta_n^{\alpha_\ell(\sigma)(\mathrm{mod}\ \ell^n)} x_n$ (for all $n \geq 0$). Then $\rho : \mathrm{Gal}(\overline{F}/F) \to \mathrm{GL}(2,\mathbb{Z}_\ell)$, $\rho : \sigma \mapsto \left(\begin{smallmatrix} \chi_\ell(\sigma) & \alpha_\ell(\sigma) \\ 0 & 1 \end{smallmatrix} \right)$, has the required properties when $\ell \neq p$. Of course, χ_ℓ is the cyclotomic character of $\mathrm{Gal}(\overline{F}/F)$, and the construction holds also in characteristic 0: if $F = \mathbb{Q}$ take p_0, $p_1,\ldots$ to be the rational primes. The semisimplification of ρ is the direct sum $1 \otimes \chi_\ell$; it is nowhere ramified.

Let $\rho : W(\overline{F}/F) \to \mathrm{Aut}\, V_\lambda$ be a λ-adic representation of the function field F; then $\rho_v = \rho|W(\overline{F}_v/F_v)$ is unramified for almost all v. Put $L(t,\rho_v) = P_{v,\rho}(t^{\deg v})^{-1}$, where $\deg v = \log_p q_v$ is the degree of the residue field of F_v over $\mathbb{F}_p$. This $L(t,\rho_v)$ is a power series in $E_\lambda[[t]]$, and so is the Euler product $L(t,\rho) = \prod_v L(t,\rho_v)$. Note that $L(t,\rho_v)$ and $L(t,\rho)$ lie in $R_\lambda[[t]]$ if ρ is a representation of $\mathrm{Gal}(\overline{F}/F)$, since then the image of ρ lies in a conjugate of $\mathrm{GL}(n, R_\lambda)$.

Proposition 12.2. (i) $L(t,\rho)$ *is a rational function in* t; (ii) *it is a polynomial in* t *if* ρ *has no nonzero* $\mathrm{Gal}(\overline{F}/F)_0$-*fixed vector;* (iii) *there exists a monomial* $\varepsilon(t,\rho)$ *in* t *such that* $L(t,\rho)$ *satisfies the functional equation* $L(t,\rho) = \varepsilon(t,\rho)L(1/qt,\rho^\vee)$, *where* $\rho^\vee$ *is the contragredient* $(\rho^\vee(w) = {}^t\rho(w)^{-1})$ *of* ρ. *The local and global L-functions satisfy* (iv) $L(t,\rho) = L(t,\rho')L(t,\rho'')$ *if* $0 \to V' \to V \to V'' \to 0$ *is an exact sequence of representations, and* (v) $L(t,\rho_F) = L(t,\rho_K)$ *if* K *is a finite extension of* F *in* $\overline{F}$, $\rho_K : W(\overline{F}/K) \to \mathrm{Aut}\, V_\lambda$ *is a* λ-*adic representation of* K, *and* ρ_F *is the induced* $\rho_F = \mathrm{Ind}(\rho_K; W(\overline{F}/F), W(\overline{F}/K))$ *representation of* F. (vi) *There exists a* $c = c(\rho) > 0$ *such that the Euler product* $L(t,\rho)$ *converges absolutely in* $|t| < c$.

Proof. This is a theorem of Grothendieck (cf. [De2], p. 574). Our (i) follows from the cohomological interpretation

$$L(t,\rho) = \prod_i \det[1 - t \cdot \mathrm{Fr} \mid H_{\text{ét}}^i(C \times \overline{\mathbb{F}}_q, \rho)]^{(-1)^{i+1}},$$

where Fr is the geometric Frobenius on the curve C which defines $F = \mathbb{F}_q(C)$, and $\overline{\mathbb{F}}_q$ is an algebraic closure of $\mathbb{F}_q$. (ii) implies the Artin conjecture for function fields, and (iii) results from the Poincaré duality. For (iv) and (v) see [De2], p. 530. (vi) follows from the interpretation of $L(t, \rho)$ as an L-function of a smooth $\overline{\mathbb{Q}}_\ell$-sheaf over some scheme; see [De2], Chap. 11, or the paragraph preceding Lemma 11.6 in the proof of Theorem 11.1, and [S1], Theorem 1. □

Remark 12.1. If ρ_v is one-dimensional, namely a character, then $L(t, \rho_v) = 1$ if ρ_v is ramified, and $L(t, \rho_v) = (1 - t \cdot \rho_v(\mathrm{Fr}_v))^{-1}$ if ρ_v is unramified. Then $L(t, \rho_v)$ coincides with the Hecke-Tate local factor for the character of $F_v^\times$ corresponding to ρ_v, as Fr_v is mapped to the uniformizer π_v by the reciprocity epimorphism of class field theory. If ρ is a character of $W(\overline{F}/F)$ then $L(t, \rho)$ coincides with the Hecke-Tate Euler product for the corresponding character of the idèle class group $\mathbb{A}^\times / F^\times$.

12.3. ε-**Factors.** To compare λ-adic representations ρ of a global field F with automorphic representations we need to express the global ε-factor as a product of local ε-factors. The local ε-factor is defined by Theorem 4.1 (and 6.5) in [De3], which we now recall.

Proposition 12.3. *There exists a unique $E_\lambda^\times$-valued function ε, associating a number $\varepsilon(\rho, \psi, \mathrm{d}x)$ to the triple consisting of (i) a λ-adic representation ρ : $W(\overline{F}/F) \to \mathrm{Aut}\, V_\lambda$ of the local field F; (ii) a nontrivial $E_\lambda^\times$-valued additive character ψ of F; and (iii) E_λ-valued (see [De2], p. 554) Haar measure $\mathrm{d}x$ on F, satisfying the following properties:*

(1) *$\varepsilon(\rho, \psi, \mathrm{d}x) = \varepsilon(\rho', \psi, \mathrm{d}x)\varepsilon(\rho'', \psi, \mathrm{d}x)$ for any exact sequence $0 \to V_\lambda' \to V_\lambda \to V_\lambda'' \to 0$ of representations. In particular, $\varepsilon(\rho, \psi, \mathrm{d}x)$ depends only on the class of (ρ, V_λ) in the Grothendieck group $R_{E_\lambda}(W(\overline{F}/F))$ of λ-adic representations of F.*

(2) *$\varepsilon(\rho, \psi, a\mathrm{d}x) = a^{\dim \rho} \varepsilon(\rho, \psi, \mathrm{d}x)$; in particular, $\varepsilon(\rho, \psi, \mathrm{d}x)$ is independent of $\mathrm{d}x$ and is denoted by $\varepsilon(\rho, \psi)$ for a virtual representation ρ of dimension zero. Note that $\dim$ and $\det$ naturally extend to homomorphisms from the Grothendieck group $R_{E_\lambda}(W(\overline{F}/F))$ to $\mathbb{Z}$ and to $\mathrm{Hom}(W(\overline{E}/E), E_\lambda^\times)$.*

(3) *If K is a finite extension of F in $\overline{F}$, ρ_K a virtual λ-adic representation of $W(\overline{F}/K)$ of dimension zero, and $\rho_F = \mathrm{Ind}(\rho_K; W(\overline{F}/F), W(\overline{F}/K))$ the induced representation of $W(\overline{F}/F)$, then $\varepsilon(\rho_F, \psi) = \varepsilon(\rho_K, \psi \circ \mathrm{tr}_{K/F})$, where $\mathrm{tr}_{K/F}$ is the trace map.*

(4) *If ρ is a character of $W(\overline{F}/F)$ corresponding to the character χ of $F^\times$ by local classified theory, then $\varepsilon(\rho, \psi, \mathrm{d}x) = \varepsilon(\chi, \psi, \mathrm{d}x)$ is the Hecke-Tate local ε-factor associated to χ (and ψ, $\mathrm{d}x$).*

For a given λ-adic representation ρ of a local field F, the ε-factor $\varepsilon(\rho \otimes \chi, \psi, \mathrm{d}x)$ depends on ρ only via its dimension and determinant, for a sufficiently ramified χ. More precisely, by [De2], we have the following:

Proposition 12.4. *For any character χ of $F^\times$ whose conductor $a(\chi)$ is sufficiently large (depending on ρ and ψ), there exists y in $F^\times$ with $\chi(1 + a) =$*

$\psi(a/y)$ for all a in $F^\times$ with $\deg(a) \geq a(\chi)/2$, and we have

$$\varepsilon(\rho \otimes \chi, \psi, dx) = \varepsilon(\chi, \psi, dx)^{\dim \rho} \cdot (\det \rho)(y).$$

Corollary 12.5. *If* $\dim \rho = r$ *and* $\chi_1, \ldots, \chi_r$ *are characters of* $W(\overline{F}/F)_{\mathrm{ab}} \simeq$ $F^\times$ *whose product is* $\det \rho$, *for a sufficiently ramified* χ, *we have*

$$\varepsilon(\rho \otimes \chi, \psi, dx) = \prod_{i=1}^{r} \varepsilon(\chi, \psi, dx)\chi_i(y) = \prod_{i=1}^{r} \varepsilon(\chi\chi_i, \psi, dx).$$

Let F be a function field, $dx = \prod_v dx_v$ a Haar measure on $\mathbb{A}$ which assigns $\mathbb{A} \bmod F$ the volume one, and ψ a nontrivial additive character of $\mathbb{A} \bmod F$. The restriction of ψ to F_v is denoted by ψ_v. Put d_v for the degree over $\mathbb{F}_p$ of the residue field of F_v, and define the unramified character $\nu_{t,v}$ of $W(\overline{F}_v/F_v)$ by $\nu_{t,v}(\mathrm{Fr}_v) = t^{d_v}$. Put

$$L(t, \rho_v) = L(\rho_v \otimes \nu_{t,v}) \quad \text{and} \quad \varepsilon(t, \rho_v, \psi_v, dx_v) = \varepsilon(\rho_v \otimes \nu_{t,v}, \psi_v, dx_v).$$

For any λ-adic (constructible) representation ρ of $W(\overline{F}/F)$, for almost all v, the factors $\varepsilon(t, \rho_v, \psi_v, dx_v)$ are equal to one, and the product

$$\varepsilon_\Pi(t, \rho) = \prod_v \varepsilon(t, \rho_v, \psi_v, dx_v)$$

is independent of the choice of ψ and decomposition $dx = \prod_v dx_v$ into local measures.

Proposition 12.6. *For every irreducible* λ-*adic representation* ρ *we have the equality* $\varepsilon(t, \rho) = \varepsilon_\Pi(t, \rho)$.

Proof. This is a main result of [Lm1].   □

Remark 12.2. The case where ρ has finite image, or it belongs to an infinite compatible system of λ-adic representations, is due to [De2], Theorem 9.3; for more historical comments see [Lm1].

Definition 12.3. Put

$$\Gamma(t, \rho_v) = \frac{L(t, \rho_v)}{\varepsilon(t, \rho_v)L(1/qt, \rho_v^\vee)}$$

and

$$\Gamma(t, \rho) = \frac{L(t, \rho)}{\varepsilon_\Pi(t, \rho)L(1/qt, \rho^\vee)}.$$

Note that $\Gamma(t, \rho_v) = \Gamma(t, \rho_v, \psi_v, dx_v)$ depends on ψ_v and dx_v.

Remark 12.3. In view of Prop. 12.6, the Grothendieck functional equation of Prop. 12.2(iii) asserts that $\Gamma(t, \rho) = 1$ for every λ-adic ρ.

12.4. Product L-Functions of Generic Representations of GL(n).

Here is a brief review of well-known results, underlying the converse theorem, and used in applications. Let $n > m > 0$ be integers. Put $k = n - m - 1$. For $v \in |X|$, let π_v and π_v' be admissible representations of $G_n = \mathrm{GL}(n, F_v)$ and $G_m = \mathrm{GL}(m, F_v)$ which are parabolically induced from irreducible generic representations. Denote by $W(\pi_v, \psi_v)$ and $W(\pi_v', \overline{\psi}_v)$ their Whittaker models. For W_v in $W(\pi_v, \psi_v)$ and W_v' in $W(\pi_v', \overline{\psi}_v)$, define

$$\Psi(t, W_v, W_v') = \int_{N_m \backslash G_m} W_v \left(\left(\begin{smallmatrix} x & 0 \\ 0 & I_{n-m} \end{smallmatrix} \right) \right) W_v'(x) (q^{\frac{n-m}{2}} t)^{\deg(v)\deg_v(x)} \, dx$$

and $\widetilde{\Psi}(t, W_v, W_v')$

$$= \int_{N_m \backslash G_m} dx \int_{\mathrm{M}(k \times m, F_v)} W_v \left(\left(\begin{smallmatrix} x & 0 & 0 \\ y & I_m & 0 \\ 0 & 0 & 1 \end{smallmatrix} \right) \right) W_v'(x) (q^{\frac{n-m}{2}} t)^{\deg(v)\deg_v(x)} \, dy,$$

where $\mathrm{M}(k \times m, F_v)$ is the space of $k \times m$ matrices. Recall that $|a_v|_v = q_v^{-\deg_v(a_v)}$, $q_v = q^{\deg(v)}$. We write $t = q^{-s}$ ($s \in \mathbb{C}$) for comparison with the characteristic 0 case. Then the factor $|\det x|_v^{s - \frac{n-m}{2}}$ becomes $(q^{\frac{n-m}{2}} t)^{\deg(v)\deg_v(x)}$, where $\deg_v(x)$ means $\deg_v(\det x)$.

Denote antidiagonal$(1, \ldots, 1)$ by w or w_n. Put $\widetilde{W}(g) = W(w \, {}^t g^{-1})$. Put $w_{n,m} = \mathrm{diag}(I_m, w_{n-m})$. The following, and the analogue for $n = m$, is due to [JPS, JS].

Proposition 12.7.   (1) *The integrals*

$$\Psi(t, W_v, W_v') \quad and \quad \widetilde{\Psi}(t, \pi_v^{\vee}(w_{n,m}) \widetilde{W}_v, \widetilde{W}_v')$$

converge absolutely in some domain $|t| < c$ to rational functions in $t^{\deg(v)}$. As W_v and W_v' range over $W(\pi_v, \psi_v)$ and $W(\pi_v', \overline{\psi}_v)$ these integrals span fractional ideals $\mathbb{C}[t, t^{-1}] L(t, \pi_v \times \pi_v')$ and $\mathbb{C}[t, t^{-1}] L(t, \pi_v^{\vee} \times \pi_v'^{\vee})$ in the ring $\mathbb{C}[t, t^{-1}]$. The L-factor $L(t, \pi_v \times \pi_v')$ is the reciprocal of a polynomial in t with constant term 1.

(2) *There exists a unique function $\varepsilon(t, \pi_v \times \pi_v', \psi_v)$ of the form $at^{m'}$ (where $a = a(\pi_v, \pi_v', \psi_v)$ in $\mathbb{C}^{\times}$, $m' = m'(\pi_v, \pi_v', \psi_v) \in \mathbb{Z}$) such that for all $W_v \in W(\pi_v, \psi_v)$ and $W_v' \in W(\pi_v', \overline{\psi}_v)$ we have*

$$\frac{\Psi(t, W_v, W_v')}{L(t, \pi_v \times \pi_v')} \varepsilon(t, \pi_v \times \pi_v', \psi_v) \omega_{\pi_v'}(-1)^m = \frac{\widetilde{\Psi}(1/qt, \pi_v^{\vee}(w_{n,m}) \widetilde{W}_v, \widetilde{W}_v')}{L(1/qt, \pi_v^{\vee} \times \pi_v'^{\vee})}.$$

(3) *There exists an integer $m(\pi_v, \pi_v', \psi_v) \geq 0$ such that for any characters χ_v, χ_v' of $F_v^{\times}$ so that $\chi_v \chi_v'$ has conductor at least $m(\pi_v, \pi_v', \psi_v)$ we have:*

(i) *The functions $\Psi(t, W_v \otimes \chi_v, W_v' \otimes \chi_v')$ and $\widetilde{\Psi}(t, \widetilde{W}_v \otimes \chi_v, \widetilde{W}_v' \otimes \chi_v')$ are polynomial in t and t^{-1}.*

(ii) $L(t, \pi_v \otimes \chi_v \times \pi_v' \otimes \chi_v') = 1$.

(iii) *If* χ_i $(1 \leq i \leq nk)$ *are characters of* $F^\times$ *whose product is equal to* $\omega_\pi^k \omega_\tau^n$, *where* ω_π *denotes the central character of* π *and* ω_τ *of* τ, *then*

$$\varepsilon(s, \pi, \tau \otimes \chi, \psi) = \prod_{i=1}^{nk} \varepsilon(s, \chi\chi_i, \psi),$$

where $\varepsilon(s, \chi, \psi)$ *is the* ε-*factor in the Hecke-Tate functional equation.*

(iv) $\varepsilon(t, \pi_v \otimes \chi_v \times \pi'_v \otimes \chi'_v, \psi_v) = \varepsilon(t, \chi_v \chi'_v, \psi_v)^{nm-1} \varepsilon(t, \chi_v \chi'_v \omega_{\pi_v}^m \omega_{\pi'_v}^n, \psi_v).$

(4) *If* π_v, π'_v *and* ψ_v *are unramified of the form* $\pi_v = I(z_i(\pi_v))$, $\pi'_v = I(z_j(\pi'_v))$ *then* $\varepsilon(t, \pi_v \times \pi'_v, \psi_v) = 1$ *and*

$$L(t, \pi_v \times \pi'_v) = \prod_{i,j}(1 - tz_i(\pi_v)z_j(\pi'_v))^{-1}, \; L(t, \pi_v^\vee \times \pi'^\vee_v)$$

$$= \prod_{i,j}(1 - t/z_i(\pi_v)z_j(\pi'_v))^{-1}.$$

Given also unramified characters χ_v *and* χ'_v *then* $(3; iv)$ *holds.*

If in addition W_v, W'_v *are the* K_v- *and* K'_v-*right invariant vectors whose value at* e *is* 1 *then* $\Psi(t, W_v, W'_v) = L(t, \pi_v \times \pi'_v)$ *and* $\widetilde{\Psi}(t, \widetilde{W}_v, \widetilde{W}'_v) = L(t, \pi_v^\vee \times \pi'^\vee_v)$.

If π_v *is also unitarizable then* $q_v^{-1/2} < |z_i(\pi_v)| < q_v^{1/2}$ *for all* i.

(5) *Suppose that* π_v, τ_v *are irreducible and* τ_v *is a cuspidal* G'_v-*module. Then* $L(s, \pi_v, \tau_v^\vee)$ *has a pole at* $s = 0$ *if and only if there is a* GL($n - n'$, F_v)-*module* τ'_v *such that* π_v *is a subquotient of* I. *Here* I *is the* G_v-*module* $I(\tau_v \times \tau'_v)$ *normalizedly induced from the representation* $\tau_v \times \tau'_v$ *of the parabolic subgroup* P *of type* $(n', n - n')$ *which is trivial on the unipotent radical of* P *and is naturally defined by* τ_v *and* τ'_v.

Let us fix a place $v \in |X|$. Denote by ν_v the character $\nu_v(x) = |x|_v = q_v^{-\deg_v(x)}$, where $q_v = q^{\deg(v)}$. If $a \in \mathbb{R}$, the unramified character ν_v^a has Hecke eigenvalue q_v^{-a}. Write $\rho(a) = \rho \otimes \nu_v^a$ if ρ is an admissible representation of GL(n, F_v). A segment is a set $\Delta = \{\rho, \rho(1), \ldots, \rho(a-1)\}$ of cuspidal (in particular irreducible) representations of GL(n, F_v), where $a \in \mathbb{Z}_{>0}$. The normalizedly parabolically induced representation $\rho \times \rho(1) \times \cdots \times \rho(a-1) = I(\rho \otimes \rho(1) \otimes \cdots \otimes \rho(a-1))$ has a unique irreducible quotient denoted $L(\Delta)$.

Two segments Δ and Δ' are called *linked* [Z] if Δ is not a subset of Δ', Δ' is not a subset of Δ, and $\Delta \cup \Delta'$ is a segment. The segment $\Delta = \{\rho, \ldots\}$ is said to *precede* $\Delta' = \{\rho', \ldots\}$ if they are linked and there is an integer $b \geq 1$ with $\rho' = \rho(b)$. The Bernstein-Zelevinski classification [Z] for GL(n, F_v) asserts

Theorem 12.8.   (1) *Suppose* $\Delta_1, \ldots, \Delta_k$ *are segments, and for all* $i < j$, Δ_i *does not precede* Δ_j. *Then the induced representation* $L(\Delta_1) \times \cdots \times$

$L(\Delta_k)$ has a unique irreducible quotient denoted $L(\Delta_1, \ldots, \Delta_k)$.

(2) The representations $L(\Delta_1, \ldots, \Delta_k)$ and $L(\Delta'_1, \ldots, \Delta'_{k'})$ are equivalent iff the sequences $\Delta_1, \ldots, \Delta_k$ and $\Delta'_1, \ldots, \Delta'_{k'}$ are equal up to order.

(3) Every admissible irreducible representation of $\mathrm{GL}(n, F_v)$ has the form $L(\Delta_1, \ldots, \Delta_k)$.

The L-factors are computed in [JPS], Sections 8 and 9:

Theorem 12.9. (1) Let $\pi_v = L(\Delta_1, \ldots, \Delta_k)$ and $\pi'_v = L(\Delta'_1, \ldots, \Delta'_{k'})$ be irreducible representations of $\mathrm{GL}(n, F_v)$ and $\mathrm{GL}(n', F_v)$. Then

$$L(t, \pi_v \times \pi'_v) = \prod_{1 \le i \le k, 1 \le i' \le k'} L(t, L(\Delta_i) \times L(\Delta'_{i'})),$$

$$L(t, \pi_v^\vee \times \pi'^\vee_v) = \prod_{1 \le i \le k, 1 \le i' \le k'} L(t, L(\Delta_i)^\vee \times L(\Delta'_{i'})^\vee).$$

(2) Let $\Delta = \{\rho, \ldots, \rho(a-1)\}$, $\Delta' = \{\rho', \ldots, \rho'(a'-1)\}$ be segments with $a' \le a$. Then

$$L(t, L(\Delta) \times L(\Delta')) = \prod_{0 \le i \le a'-1} L(t, \rho(a-1) \times \rho'(i)),$$

$$L(t, L(\Delta)^\vee \times L(\Delta')^\vee) = \prod_{0 \le i \le a'-1} L(t, \rho(a-1)^\vee \times \rho'(-i)^\vee).$$

(3) Let ρ, ρ' be cuspidal representations of $\mathrm{GL}(n, F_v)$ and $\mathrm{GL}(n', F_v)$. Then

$$L(t, \rho \times \rho') = \prod_z (1 - z^{-1} t^{\deg(v)})^{-1}, \quad L(t, \rho^\vee \times \rho'^\vee) = \prod_z (1 - zt^{\deg(v)})^{-1},$$

where z ranges over the numbers for which $\rho \otimes z^{\deg_v}$ is equivalent to $\rho'^\vee$ (and $\deg_v(h) = \deg_v(\det h)$).

Let π_v be an admissible representation of $\mathrm{GL}(n, F_v)$ with central character ω_{π_v}. Denote by $|\pi_v|$ the unique unramified character of $\mathrm{GL}(n, F_v)$ into $\mathbb{R}^\times_{>0}$ such that the central character of $\pi_v \otimes |\pi_v|^{-1}$ is unitary. Thus $|\pi_v| = |\omega_{\pi_v}|^{1/n}$. If ρ is cuspidal then $\rho \otimes |\rho|^{-1}$ is unitarizable. For a segment $\Delta = \{\rho, \ldots, \rho(a-1)\}$, put $|\Delta| = |\rho(\frac{a-1}{2})|$. Then $|\Delta| = |L(\Delta)|$, and $L(\Delta) \otimes |\Delta|^{-1}$ is unitarizable and square integrable. The representation $L(\Delta_1, \ldots, \Delta_k)$ is tempered when $|\Delta_i| = 1$ ($1 \le i \le k$). It is then equal to $L(\Delta_1) \times \cdots \times L(\Delta_k)$, which is irreducible and unitarizable. An extension of the last theorem for unitarizable π_v by Tadic asserts

Theorem 12.10. Let $\pi_v = L(\Delta_1, \ldots, \Delta_k)$ be an irreducible unitarizable generic representation of $\mathrm{GL}(n, F_v)$. Then the Hecke eigenvalues $z(|\Delta_i|) = |\Delta_i|(\pi_v)$ of the unramified character $|\Delta_i|$ satisfy $q_v^{-1/2} < z(|\Delta_i|) < q_v^{1/2}$ ($1 \le i \le k$).

Corollary 12.11. *Let π_v, π'_v be admissible irreducible generic representations of $GL(n, F_v)$ and $GL(n', F_v)$.*

(1) *If both are tempered, then the poles of $L(t, \pi_v \times \pi'_v)$ and $L(t, \pi_v^\vee \times \pi'^\vee_v)$ are in $|t| \geq 1$.*

(2) *If one is tempered and the other is unitarizable then these poles are in $|t| > q^{-1/2}$.*

Other well-known results of [CPS, JPS, JS] include

Theorem 12.12. *Let $\pi = \otimes_v \pi_v$, $\pi' = \otimes_v \pi'_v$ be irreducible automorphic representations of $GL(n, \mathbb{A})$ and $GL(n', \mathbb{A})$. Then*

(1) *The global ε-factor $\varepsilon(t, \pi \times \pi') = \prod_v \varepsilon_v(t, \pi_v \times \pi'_v, \psi_v)$ is independent of ψ.*

(2) *The power series $L(t, \pi \times \pi') = \prod_v L(t, \pi_v \times \pi'_v)$ (and consequently $L(t, \pi^\vee \times \pi'^\vee)$) converge to a rational function in t in some domain.*

(3) *These rational functions satisfy the functional equation*

$$L(t, \pi \times \pi') = \varepsilon(t, \pi \times \pi') L(1/qt, \pi^\vee \times \pi'^\vee).$$

(4) *If π is cuspidal and $n' < n$ then the rational functions $L(t, \pi \times \pi')$ and $L(t, \pi^\vee \times \pi'^\vee)$ are polynomials.*

From [JS], II, (3.7) we have

Theorem 12.13. *Let $\pi = \otimes_v \pi_v$, $\pi' = \otimes_v \pi'_v$ be cuspidal (unitary, irreducible) representations of $GL(n, \mathbb{A})$ and $GL(n', \mathbb{A})$. Let S be a finite subset of $|X|$ such that π_v, π'_v are unramified at each $v \notin S$. Then on $|t| \leq q^{-1}$ the partial L-function $L^S(t, \pi \times \pi'^\vee)$ has no zeroes, and its poles are all simple. The poles z in $|z| \leq q^{-1}$ are the numbers z with $|z| = q^{-1}$ such that $(n = n'$ and$)$ π is equivalent to $\pi' \otimes (qz)^{\deg(.)}$.*

There is a unique way to complete a partial L-function to an L-function.

Proposition 12.14. *Suppose for each place $v \in |X|$ and each pair χ_v, χ'_v of characters of finite order of $F_v^\times$ we have two triples*

$$L^i(t, \chi_v, \chi'_v), \quad L^{\vee i}(t, \chi_v, \chi'_v), \quad \varepsilon^i(t, \chi_v, \chi'_v, \psi_v) \quad (i = 1, 2)$$

such that

(1) *for some finite $S \subset |X|$ the triples are independent of $i = 1, 2$ for all $v \notin S$ (and all χ_v, χ'_v);*

(2) *for all $v \in S$ the same holds provided $\chi_v \chi'_v$ is sufficiently ramified. Suppose*

(3) *for each pair* $\chi = \otimes_v \chi_v$, $\chi' = \otimes_v \chi'_v$ *of characters of finite order of* $\mathbb{A}^\times / F^\times$ *the formal products*

$$L^i(t, \chi, \chi') = \prod_v L^i(t, \chi_v, \chi'_v),$$

$$L^{\vee i}(t, \chi, \chi') = \prod_v L^{\vee i}(t, \chi_v, \chi'_v) \quad (i = 1, 2)$$

define rational functions, the products $\varepsilon^i(t, \chi, \chi') = \prod_v \varepsilon^i(t, \chi_v, \chi'_v, \psi_v)$ *are finite (almost all factors are 1), and the functional equations*

$$L^i(t, \chi, \chi') = \varepsilon^i(t, \chi, \chi') L^{\vee i}(1/qt, \chi, \chi')$$

hold. Then $\varepsilon^i(t, \chi_v, \chi'_v, \psi_v) L^{\vee i}(t, \chi_v, \chi'_v) / L^i(t, \chi_v, \chi'_v)$ *is independent of i for every v. Moreover, if the rational functions* $L^i(t, \chi_v, \chi'_v)$ *and* $L^{\vee i}(t, \chi_v, \chi'_v)$ *do not have a common pole ($i = 1, 2$), then* $L^i(t, \chi_v, \chi'_v)$, $L^{\vee i}(t, \chi_v, \chi'_v)$, *and* $\varepsilon^i(t, \chi_v, \chi'_v, \psi_v)$ *are independent of i.*

Proof. It suffices to show the claim for $v \in S$, by (1). Fix characters of finite order χ_v, χ'_v. At each $u \neq v$ in S fix characters χ_u, χ'_u of finite order with $\chi_u \chi'_u$ sufficiently ramified for (2) to hold. Let χ, χ' be characters of $\mathbb{A}^\times / F^\times$ whose components at v and $u \in S$ are those fixed. Then claim then follows from (3). The "Moreover" follows from the form of the L and ε. $\square$

Proposition 12.14 applies in particular when $\pi = \otimes_v \pi_v$, $\pi' = \otimes_v \pi'_v$ are automorphic irreducible representations of $\mathrm{GL}(n, \mathbb{A})$ and $\mathrm{GL}(n', \mathbb{A})$, and

$$L^1(t, \chi_v, \chi'_v) = L(t, \chi_v \pi_v \times \chi'_v \pi'_v), \quad L^{\vee 1}(t, \chi_v, \chi'_v) = L(t, \chi_v^{-1} \pi_v^\vee \times \chi'_v{}^{-1} \pi'_v{}^\vee),$$

and $\varepsilon^1(t, \chi_v, \chi'_v) = \varepsilon(t, \chi_v \pi_v \times \chi'_v \pi'_v, \psi_v)$. The assumptions (1), (2), and (3) follow from Props. 12.7(4) and 12.7(3) and Theorem 12.12. The assumption of "Moreover" is satisfied by Corollary 12.11 when π_v, π'_v are generic and both are tempered or one is tempered and the other unitary.

12.5. Correspondence.

Section 12.4 completes our summary of the theory of λ-adic representations ρ of the Weil group $W(\overline{F}/F)$ of a local or global field of characteristic p. These ρ will be related now to cuspidal representations of $G = \mathrm{GL}(r)$. As usual, $\mathbb{A}$ denotes the ring of adèles of a function field F of characteristic p, and each irreducible $G(\mathbb{A})$-module π is the restricted product $\otimes_v \pi_v$ of irreducible $G_v = G(F_v)$-modules π_v which are almost all unramified. Denote now by C the field $\mathbb{C}$ or $\overline{\mathbb{Q}}_\ell$ ($\ell \neq p$). Suppose that π and π_v are realizable in a vector space over C.

If π_v is an irreducible unramified G_v-module in a space over the field C, where G_v is $\mathrm{GL}(r, F_v)$, then there exists an r-tuple $z(\pi_v) = (z_i)$ of nonzero elements z_i of C such that π_v is the unique irreducible unramified subquotient of the G_v-module $I(z(\pi_v); G_v, B_v) = \mathrm{Ind}(\delta^{1/2} z(\pi_v); G_v, B_v)$ normalizedly induced from the unramified character $z(\pi_v) : (b_{ij}) \mapsto \prod_i z_i^{\deg_v(b_{ii})}$ of the upper triangular subgroup B_v. Here $\deg_v$ is the normalized (integral valued) additive valuation on $F_v^\times$, and

$$\delta((b_{ij})) = \prod_i q_v^{\deg_v(b_{ii})(i-(r+1)/2)} = \prod_i |b_{ii}|_v^{(r+1)/2-i}.$$

Recall that $|x|_v = q_v^{-\deg_v(x)}$ and q_v is the cardinality of the residual field of F_v. The polynomial $P_{v,\pi}(t) = \prod_i(1 - tz_i)$ is uniquely determined by π_v, and $P_{v,\pi}(t) \neq P_{v,\pi'}(t)$ if π_v, π_v' are irreducible unramified inequivalent G_v-modules.

Denote by Z the center of $G = \mathrm{GL}(r)$. Let ω be a unitary character of $Z(\mathbb{A})/Z$. Here we write Z for $Z(F)$. Let $L(G)_\omega$ be the space of C-valued functions φ on $G(\mathbb{A})$ such that there is an open subgroup $U = U_\varphi$ of $G(\mathbb{A})$ with $\varphi(z\gamma gu) = \omega(z)\varphi(g)$ for all z in $Z(\mathbb{A})$, γ in G, g in $G(\mathbb{A})$, and u in U_φ. Let $L^2(G)_\omega$ be the subspace of φ with $\int_{Z(\mathbb{A})G\backslash G(\mathbb{A})} |\varphi(g)|^2 dg < \infty$. Let $L_0(G)_\omega$ be the subspace of *cuspidal* functions φ in $L(G)_\omega$, those with $\int_{N_P\backslash N_P(\mathbb{A})} \varphi(ng)dn = 0$ for every proper parabolic subgroup P of G; here N_P denotes the unipotent radical of P. Then $L_0(G)_\omega$ is contained in $L^2(G)_\omega$. In fact, by a theorem of Harder, cuspidal φ are compactly supported modulo $Z(\mathbb{A})$. An admissible $G(\mathbb{A})$-module $\pi = \otimes_v\pi_v$ with central character ω is called *cuspidal* if it is a constituent, necessarily a direct summand, of the representation of $G(\mathbb{A})$ on $L_0(G)_\omega$ by right translation. It is called here *automorphic* if it is a constituent of $L(G)_\omega$.

Definition 12.4. Let F be a function field, ρ a finite dimensional representation of $W(\overline{F}/F)$ over E_λ, and π an admissible irreducible $G(\mathbb{A})$-module over E_λ. Then π and ρ are called *corresponding* if $P_{v,\rho}(t) = P_{v,\pi}(t)$ for almost all v.

Remark 12.4.   (1) If ρ corresponds to π write $\pi = \pi(\rho)$ and $\rho = \rho(\pi)$. By definition such ρ has dimension r and it is constructible. Moreover, the central character ω of π corresponds to the determinant $\det \rho$ of ρ under the isomorphism $\mathbb{A}^\times/F^\times \simeq W(\overline{F}/F)_{\mathrm{ab}}$.

(2) If ρ_i corresponds to π_i ($1 \leq i \leq j$) then $\oplus_i\rho_i$ corresponds to any irreducible constituent of the $G(\mathbb{A})$-module $I = I(\otimes_i\pi_i)$ normalizedly induced from the $P(\mathbb{A})$-module $\otimes_i\pi_i$ which is trivial on the unipotent radical N of $P(\mathbb{A})$; here P is the standard parabolic subgroup with Levi subgroup $\prod_i \mathrm{GL}(r_i)$ if $\dim \rho_i = r_i$. Note that this definition is compatible with that of the local correspondence given below in terms of local L and ε-factors, only when I is irreducible.

(3) If ρ corresponds to π and χ is a character of $W(\overline{F}/F)_{\mathrm{ab}} \simeq \mathbb{A}^\times/F^\times$, then $\rho \otimes \chi$ corresponds to $\pi \otimes \chi$, and $\rho^\vee$ corresponds to $\pi^\vee$ where $\rho^\vee$, $\pi^\vee$ are the representations contragredient to ρ and π.

(4) We defined the correspondence using the geometric Frobenius, as is usually done. Defining it using the arithmetic Frobenius, the representation ρ would be replaced by its contragredient.

Proposition 12.15. *Let ρ be an irreducible r-dimensional λ-adic representation of $W(\overline{F}/F)$ whose determinant $\det \rho$ is of finite order. Then ρ extends to a representation of $\mathrm{Gal}(\overline{F}/F)$.*

Proof. (cf. [De2], Section 4.10). Let ρ_0 denote an irreducible constituent in the restriction of ρ to $\mathrm{Gal}(\overline{F}/F)_0$. Let W_0 denote the group of w in $W(\overline{F}/F)$ with $\rho_0^w \simeq \rho_0$; here we put $\rho_0^w(g) = \rho_0(w^{-1}gw)$. Then ρ_0 extends to a representation of W_0, and $\rho = \mathrm{Ind}(\rho_0; W(\overline{F}/F), W_0)$. Since ρ has a finite dimension r, the index m of W_0 in $W(\overline{F}/F)$ is finite. Let Fr denote an element of $W(\overline{F}/F)$ whose image in $\mathbb{Z}$ is 1. Then Fr^m lies in W_0, and $\rho(\mathrm{Fr}^m)$ is a scalar, since ρ is irreducible. Suppose that $\det \rho$ has order k. Then $1 = \det \rho(\mathrm{Fr}^m)^k = \rho(\mathrm{Fr}^{mrk})$. Since the image of $\mathrm{Gal}(\overline{F}/F)_0$ under ρ is profinite, so is the image of $W(\overline{F}/F)$, being a finite extension of $\rho(\mathrm{Gal}(\overline{F}/F)_0)$. But $\mathrm{Gal}(\overline{F}/F)$ is the profinite completion of $W(\overline{F}/F)$. Hence ρ extends to a representation of $\mathrm{Gal}(\overline{F}/F)$, as required. $\qquad\square$

Corollary 12.16. *If ρ is an irreducible λ-adic representation of $W(\overline{F}/F)$ then there exists a character χ of $W(\overline{F}/F)$ such that $\rho \otimes \chi$ extends to a representation of $\mathrm{Gal}(\overline{F}/F)$.*

Proof. Given a character χ of $W(\overline{F}/F)$ there exists a character χ' of $W(\overline{F}/F)$ which is trivial on $\mathrm{Gal}(\overline{F}/F)_0$ such that $\chi\chi'$ is of finite order. Indeed, χ can be viewed as a character of $\mathbb{A}^\times/F^\times$ by class field theory. Moreover, χ' can be taken to factorize via the volume character $x \mapsto |x|$ of $\mathbb{A}^\times$, since the restriction of χ to the group $\mathbb{A}^0$ of idèles of volume one has finite order. $\qquad\square$

Let ∞ be a fixed place of F. Let E_λ be a finite extension of $\mathbb{Q}_\ell$, where $\ell \neq p$ is a rational prime. We recall Corollary 11.2 as

Theorem 12.17. *For any irreducible cuspidal λ-adic representation $\pi = \otimes_v \pi_v$ of $G(\mathbb{A})$ whose component π_∞ is cuspidal, there exists a unique irreducible λ-adic r-dimensional representation ρ of $W(\overline{F}/F)$ which corresponds to π.*

Our subsequent results in Sects. 12.5 and 12.6 depend on this theorem, which relies on Deligne's conjecture (Theorem 6.8). The main application is the following global higher reciprocity law relating cuspidal and irreducible λ-adic representations.

Theorem 12.18. *The correspondence defines a bijection between the sets of equivalence classes of irreducible (1) cuspidal $G(\mathbb{A})$-modules π whose component π_∞ at ∞ is cuspidal and (2) r-dimensional continuous ℓ-adic constructible representations ρ of $W(\overline{F}/F)$ whose restriction ρ_∞ to $W(\overline{F}_\infty/F_\infty)$ is irreducible. The determinant $\det \rho$ of ρ corresponds by class field theory to the central character of π.*

Remark 12.5. By virtue of the Chebotarev density theorem (see, e.g., [S2]), the irreducible ρ is uniquely determined by its restriction to $W(\overline{F}_{\overline{v}}/F_v)$ for almost all v. The rigidity theorem (aka strong multiplicity one theorem, see [JS]) for $GL(r)$ asserts that the cuspidal π is uniquely determined by the set of its components π_v for almost all v. Hence the uniqueness assertion of the reciprocity law is clear; the existence is to be proven.

This global reciprocity law will be accompanied by its local analogue, the local reciprocity law for representations of G_v and $W(\overline{F}_{\overline{v}}/F_v)$. We put $t = q^{-s}$.

Theorem 12.19. *For every local field F_v of characteristic $p > 0$, and for every $r \geq 1$ there is a unique bijection $\pi_v \leftrightarrow \rho_v$ between the sets of equivalence classes of irreducible (1) cuspidal G_v-modules π_v and (2) continuous ℓ-adic r-dimensional representations ρ_v of $W(\overline{F_v}/F_v)$, with the following properties. (i) If $\pi_v \leftrightarrow \rho_v$ then (1) $\pi_v \otimes \chi_v \leftrightarrow \rho_v \otimes \chi_v$ for every character χ_v of $F_v^\times \simeq W(\overline{F_v}/F_v)_{\mathrm{ab}}$; (2) the central character of π_v corresponds to $\det \rho_v$ by local class field theory; (3) the contragredient of π_v corresponds to the contragredient of ρ_v. (ii) If the $\mathrm{GL}(n, F_v)$-module $\pi_v^{(n)}$ corresponds to $\rho_v^{(n)}$, and the $\mathrm{GL}(m, F_v)$-module $\pi_v^{(m)}$ corresponds to $\rho_v^{(m)}$, then*

$$\Gamma(t, \pi_v^{(m)}, \pi_v^{(n)}) = \Gamma(t, \rho_v^{(m)} \otimes \rho_v^{(n)})$$

for all ψ_v, $\mathrm{d}x_v$. Moreover, this bijection has the property that π and ρ correspond by Theorem 12.10 if and only if π_v and ρ_v correspond for all v.

Remark 12.6. By virtue of [Z], Section 10, there is a unique natural extension of this local correspondence to relate the sets of equivalence classes of (A) irreducible G_v-modules π_v and (B) continuous ℓ-adic r-dimensional representations ρ_v of $W(\overline{F}_v/F_v)$, which satisfies (i), commutes with induction, and bijects square integrable π_v with indecomposable ρ_v.

Lemma 12.20. (1) *Every local cuspidal representation is a component of a global cuspidal representation.* (2) *Every irreducible local λ-adic representation with finite image is the restriction of a global λ-adic representation.*

Proof.   (1) Given a local field F_w there exists a global field F whose completion is F_w. Given a cuspidal G_w-module π_w^0 there exists a cuspidal $G(\mathbb{A})$-module π whose component at w is π_w^0. This is easily seen by means of the trace formula with a test function f whose component at w is a matrix coefficient f_w of π_w^0, which is nonzero on a single $G(\mathbb{A})$-orbit of an elliptic regular conjugacy class in $G(F)$; see, e.g., [F2], III. In fact, for each place $u \neq w$ and square-integrable G_{v_i}-modules $\pi_{v_i}^0$ $(1 \leq i \leq m;\ v_i \neq u, w)$, π can be taken to have the components $\pi_{v_i}^0$, in addition to π_w^0, and its components π_v for $v \neq u, w, v_i$ can be taken to be unramified. Moreover, m can be taken to be 0.

(2) If E_w/F_w is a finite Galois extension of local fields then there exists a finite Galois extension E/F of global fields such that F_w is the completion of F at a place w and $E_w = E \otimes_F F_w$, and $\mathrm{Gal}(E/F) \simeq \mathrm{Gal}(E_w/F_w)$. Consequently if ρ_w is an irreducible representation of $\mathrm{Gal}(E_w/F_w)$, then there exists a representation ρ of $\mathrm{Gal}(E/F)$ whose restriction to the decomposition subgroup $\mathrm{Gal}(E_w/F_w)$ is ρ_w. $\square$

12.6. Smooth Sheaves. We continue with a more detailed description of the correspondence, in the equivalent language of ℓ-adic sheaves. As usual ℓ denotes a rational prime number prime to the cardinality q of the base field $\mathbb{F}_q$. Let C be a curve over $\mathbb{F}_q$ or, more generally, a scheme of finite type.

Denote by $S_\ell(C)$ the set of isomorphism classes of smooth ℓ-adic sheaves on C. If C is connected and $\overline{v}$ is a geometric point of C, then $S_\ell(C)$ is isomorphic to the set of isomorphism classes of continuous finite dimensional representations of the fundamental group $\pi_1(C, \overline{v})$ over a finite extension E_λ of $\mathbb{Q}_\ell$, see [DF].

In particular each smooth ℓ-adic sheaf on C has constant rank equal to the dimension of the corresponding representation of $\pi_1(C, \overline{v})$. Denote by $S_\ell^r(C)$ the subset of isomorphism classes of smooth ℓ-adic sheaves on C of rank r.

There is a canonical continuous surjective homomorphism $\pi_1(C, \overline{v}) \to \mathrm{Gal}(\overline{\mathbb{F}}_q / \mathbb{F}_q) \simeq \widehat{\mathbb{Z}}$, where the Galois group of $\mathbb{F}_q$ is generated by the Frobenius substitution $\varphi : x \mapsto x^q$. The inverse image of $\mathbb{Z}$ is the Weil group $W(C, \overline{v})$. It is a dense subgroup of $\pi_1(C, \overline{v})$. Every ℓ-adic representation of $\pi_1(C, \overline{v})$ is uniquely determined by its restriction to $W(C, \overline{v})$. Denote the geometric Frobenius automorphism by $\mathrm{Fr} = \varphi^{-1}$.

Let $|C|$ denote the set of closed points of C. Let $\rho \in S_\ell(C)$ be a smooth ℓ-adic sheaf of rank r on C. The fiber ρ_v of ρ at a closed point v in $|C|$ can be viewed as an r-dimensional vector space over E_λ with an action of the Frobenius $\mathrm{Fr}_v = \mathrm{Fr}^{\deg(v)}$. Denote by $z_1(\rho_v), \ldots, z_r(\rho_v)$ the eigenvalues of the (arithmetic) Frobenius substitution and by $u_i(\rho_v) = z_i(\rho_v)^{-1}$ those of the (geometric) Frobenius.

Define the local L-factor of ρ at v to be

$$L(t, \rho_v) = \det(I - t^{\deg(v)} \, \mathrm{Fr}_v \, |\rho_v)^{-1} = \prod_{1 \le i \le r} (1 - z_i(\rho_v)^{-1} t^{\deg(v)})^{-1}.$$

Define the global L-function of ρ to be $L(t, \rho, C) = \prod_{v \in |C|} L(t, \rho_v)$. It is a formal power series in t with coefficients in E_λ.

More generally one defines $L(t, \rho_v)$ and $L(t, \rho, C)$ in the same way when ρ is a constructible ℓ-adic sheaf on C. The definition can be made also when C denotes any connected scheme of finite type over $\mathbb{F}_q$.

Recall Grothendieck's

Theorem 12.21. *Let ρ be a constructible (in particular smooth) ℓ-adic sheaf on a scheme C of finite type over $\mathbb{F}_q$. Let $H_c^i(\rho)$ ($0 \le i \le 2 \dim C$) be the étale cohomology spaces with compact support of ρ over $\mathrm{Spec}\,\mathbb{F}_q$. Then the power series $L(t, \rho, C)$ is equal to the rational function in t:*

$$\prod_{0 \le i \le 2 \dim C} [\det(I - t \cdot \mathrm{Fr} \, |H_c^i(\rho))]^{(-1)^{i+1}}.$$

Fix as usual an isomorphism of $\overline{\mathbb{Q}}_\ell$ with $\mathbb{C}$.

A smooth sheaf $\rho \in S_\ell(C)$ is called *pure of weight n* if for all $v \in |C|$ the eigenvalues of the geometric Frobenius Fr_v in the fiber ρ_v have absolute value $q^{n \deg(v)/2}$.

We say that ρ is *mixed of weight $\le n$* if it has a filtration whose successive quotients are pure of weights $\le n$. Recall Deligne's

Theorem 12.22. *If ρ is mixed of weight $\leq n$ then $H_c^i(\rho)$ is mixed of weight $\leq n + i$, for all $i \geq 1$. If C is smooth and proper over $\mathbb{F}_q$, and ρ is pure of weight n, then $H_c^i(\rho)$ is pure of weight $n + i$, for all $i \geq 1$.*

Suppose $s \in \mathbb{C}$ is such that $q^s \in \mathbb{C} \simeq \overline{\mathbb{Q}}_\ell$ is an ℓ-adic unit. Then it defines a smooth ℓ-adic sheaf $\mathbb{Q}_\ell(s)$ of rank one over $\operatorname{Spec} \mathbb{F}_q$ and by pullback, on any scheme C of finite type over $\mathbb{F}_q$. If $\rho \in S_\ell(C)$ then $\rho(s) = \rho \otimes_{\mathbb{Q}_\ell} \mathbb{Q}_\ell(s)$ lies in $S_\ell(C)$. For any $s \in C$ we have that $\rho(s)$ is a smooth ℓ-adic sheaf on $C \otimes_{\mathbb{F}_q} \overline{\mathbb{F}}_q$ with an action of Fr, that is, a representation of the Weil group $W(C, \overline{v})$ at any geometric point $\overline{v}$ of C.

In view of Theorems 12.21 and 12.22, the following follows from the fact that $H_c^{2\dim C}(\rho \otimes \rho'^\vee \otimes \mathbb{Q}_\ell(\dim C + s))$ is the dual of $\operatorname{Hom}(\rho'(-s), \rho)$.

Corollary 12.23. *Let C be a scheme of finite type and geometrically connected over $\mathbb{F}_q$. Let $\rho \in S_\ell(C)$ be a mixed ℓ-adic sheaf of weight $\leq n$ and $\rho' \in S_\ell(C)$ an irreducible pure ℓ-adic sheaf of weight $\leq m$. Then $L(t, \rho \otimes \rho'^\vee, C)$ has no zero in $|t| < q^{\frac{1}{2}(m-n+1)-\deg C}$. It has poles precisely at the points of the form $q^{-\dim C - s}$ with $\operatorname{Re} s = \frac{1}{2}(n - m)$ such that $\rho'(-s)$ is a subsheaf of ρ. The order of such a pole is equal to the multiplicity of $\rho'(-s)$ in ρ.*

To deal with the ramified places, we now take C to be a smooth projective geometrically connected curve over $\mathbb{F}_q$. As usual, F denotes its function field, F_v the completion at a closed point $v \in |C|$, O_v the ring of integers in F_v and $\mathbb{F}_v$ the residue field. A smooth ℓ-adic sheaf ρ_v on $\operatorname{Spec} F_v$ can be viewed as an ℓ-adic representation of the Galois group $\operatorname{Gal}(\overline{F}_v/F_v)$ of F_v. Its direct image under the open immersion $\operatorname{Spec} F_v \hookrightarrow \operatorname{Spec} O_v$ is a constructible ℓ-adic sheaf. Denote by $\overline{\rho}_v$ its fiber at the closed point v. It can be viewed as a vector space over E_λ with action of $\operatorname{Gal}(\overline{\mathbb{F}}_v/\mathbb{F}_v)$, that is, with an action of the Frobenius $\operatorname{Fr}_v = \operatorname{Fr}^{\deg(v)}$. The local L-factor of ρ_v is

$$L(t, \rho_v) = \det[I - t^{\deg(v)} \cdot \operatorname{Fr}_v |\overline{\rho}_v]^{-1}.$$

We say that ρ_v is *unramified* if it extends to a smooth ℓ-adic sheaf on $\operatorname{Spec} O_v$. An ℓ-adic sheaf χ_v on $\operatorname{Spec} F_v$ is *invertible* if it is a character of the Galois group of F_v. From the definition of $L(t, \rho_v)$ we obtain

Lemma 12.24. *Let ρ_v be a smooth ℓ-adic sheaf over $\operatorname{Spec} F_v$. Let χ_v be an invertible ℓ-adic sheaf on $\operatorname{Spec} F_v$. If ρ_v and χ_v are unramified then*

$$L(t, \rho_v \otimes \chi_v) = \det[I - t^{\deg(v)} \cdot \operatorname{Fr}_v |\rho_v \otimes \chi_v]^{-1}.$$

If ρ_v is unramified but χ_v is ramified, and more generally if χ_v is sufficiently ramified with respect to ρ_v, then $L(t, \rho_v \otimes \chi_v) = 1$.

Denote now by $S_\ell(F)$ the set of isomorphism classes of smooth ℓ-adic sheaves ρ on $\operatorname{Spec} F$ which extend to a smooth sheaf on an open subscheme of the curve C. This is the set of isomorphism classes of smooth ℓ-adic sheaves ρ on $\operatorname{Spec} F$ whose direct image under $\operatorname{Spec} F \hookrightarrow C$ is a constructible sheaf on C which is smooth on a nonempty open subset. Such a ρ defines a smooth sheaf on $\operatorname{Spec} F_v$. The fiber of the constructible sheaf ρ at each closed point $v \in |C|$ is equal to $\overline{\rho}_v$. We obtain the local L-factors $L(t, \rho_v)$ at all closed points $v \in |C|$. Theorems 12.21 and 12.22 imply

Proposition 12.25. *Let* $\rho \in S_\ell(F)$ *be an ℓ-adic sheaf which is smooth and pure of weight* $n \in \mathbb{Z}$ *on a nonempty open subset of the curve* C. *Then at each closed point* $v \in |C|$ *each pole of the rational function* $L(t, \rho_v)$ *has absolute value of the form* $q^{-\frac{1}{2}(n-m)}$ *for some integer* $m \geq 0$.

Given ρ in $S_\ell(F)$ we now have the global L-function $L(t, \rho) = \prod_{v \in |C|} L(t, \rho_v)$ on the entire curve C. It is a power series in t, which is a rational function by Grothendieck's Theorem 12.21. Denote by $\rho^\vee$ the dual in $S_\ell(F)$ of ρ. The functional equation of Grothendieck, derived from Poincaré duality, relates their L-functions (cf. [Lf2], Théorème VI.6, pp. 155–156).

Theorem 12.26. *We have* $L(t, \rho) = \varepsilon(t, \rho) L(1/qt, \rho^\vee)$ *for all ℓ-adic sheaves* ρ *in* $S_\ell(F)$, *where* $\varepsilon(t, \rho) = \prod_{0 \leq i \leq 2} \det[-t \cdot \mathrm{Fr} \,| H_c^i(\rho)]^{(-1)^{i+1}}$ *is the product of a nonzero constant and a power of* t.

There is a product formula for $\varepsilon(t, \rho)$, due to Langlands, Deligne, and Laumon, recorded already in Prop. 12.6, which we repeat in Theorem 12.28.

Let ψ be a nontrivial additive character of $\mathbb{A} \bmod F$, where $\mathbb{A}$ denotes the ring of adèles of F. Its restriction to $F_v \hookrightarrow \mathbb{A}$ ($v \in |C|$) is a nontrivial additive character of F_v. The choice of ψ amounts to a choice of a nontrivial meromorphic differential form on the curve C. As in [De2], [Lm1, 3.1.5], there is a local factor $\varepsilon(t, \rho_v, \psi_v)$ for each closed point v of C and a smooth ℓ-adic sheaf ρ_v on $\mathrm{Spec}\, F_v$. It is a product of a nonzero constant and a power of t. In some simple cases it is given by

Lemma 12.27. *Let* ρ_v *be a smooth ℓ-adic sheaf of rank* r *and* χ_v *an invertible ℓ-adic sheaf on* $\mathrm{Spec}\, F_v$, $v \in |C|$. *If* ρ_v *is unramified then* $\varepsilon(t, \rho_v \otimes \chi_v, \psi_v) = \varepsilon(t, \chi_v, \psi_v)^{r-1} \varepsilon(t, \det(\rho_v) \otimes \chi_v, \psi_v)$. *It is equal to 1 if* χ_v *and* ψ_v *are also unramified. The same equality holds for any* ρ_v *provided* χ_v *is sufficiently ramified as a function of* ρ_v.

Then for $\rho \in S_\ell(F)$, the local ε-factors $\varepsilon(t, \rho_v, \psi_v)$ are 1 for all $v \in |C|$ except for finitely many v, so their product is well defined. As a consequence of his theory of Fourier transform, Laumon [Lm1] proved

Theorem 12.28. *We have* $\varepsilon(t, \rho) = \prod_{v \in |C|} \varepsilon(t, \rho_v, \psi_v)$ *for all ℓ-adic sheaves* ρ *in* $S_\ell(F)$.

Let ∞ denote a fixed closed point of C. Denote by $A^r(F, \infty)$ the set of irreducible cuspidal representations π of $G(\mathbb{A})$ whose component π_∞ at ∞ is cuspidal. Denote by $S_\ell^r(F, \infty)$ the set of equivalence classes of irreducible smooth ℓ-adic sheaves ρ on $\mathrm{Spec}\, F$, whose direct image under the morphism $\mathrm{Spec}\, F \to C$ is a constructible sheaf on C which is smooth on an open subset V of the curve C, whose rank (dimension of the associated representation of $\pi_1(V, \overline{v})$) is r, and the deduced smooth ℓ-adic sheaf on $\mathrm{Spec}\, F_\infty$ (representation of $\mathrm{Gal}(\overline{F}_\infty/F_\infty)$) is irreducible.

We next conclude from Theorem 12.17 (or 11.1) the following local

Theorem 12.29. *Given* $\pi \in A^r(F, \infty)$ *and* $\pi' \in A^{r'}(F, \infty)$, *let* $\rho \in S_\ell^r(F, \infty)$ *and* $\rho' \in S_\ell^{r'}(F, \infty)$ *be the corresponding ℓ-adic sheaves. Then for each $v \in |C|$ we have*

$$L(t, \pi_v \times \pi'_v) = L(t, \rho_v \times \rho'_v), \qquad \varepsilon(t, \pi_v \times \pi'_v, \psi_v) = \varepsilon(t, \rho_v \times \rho'_v, \psi_v).$$

Proof. Let S denote the set of places $v \in |C|$ where π_v or π'_v, hence also ρ_v or ρ'_v, are ramified. It follows from Theorem 12.17 and Lemmas 12.24 and 12.27 that for any characters χ, χ' of $\mathbb{A}^\times / F^\times \simeq W(\overline{F}/F)_{\mathrm{ab}}$ of finite order, at each place $v \notin S$, we have

$$L(t, \chi_v \pi_v \times \chi'_v \pi'_v) = L(t, \rho_v \otimes \rho'_v \otimes \chi_v \chi'_v),$$

$$L(t, \chi_v^{-1} \pi_v^\vee \times \chi_v'^{-1} \pi_v'^\vee) = L(t, \rho_v^\vee \otimes \rho_v'^\vee \otimes \chi_v^{-1} \chi_v'^{-1}),$$

$$\varepsilon(t, \chi_v \pi_v \times \chi'_v \pi'_v, \psi_v) = \varepsilon(t, \rho_v \otimes \rho'_v \otimes \chi_v \chi'_v, \psi_v).$$

By the same references, these equalities hold at each place $v \in S$ as long as $\chi_v \chi'_v$ is sufficiently ramified.

The products of these local factors satisfy the functional equations

$$L(t, \chi\pi \times \chi'\pi') = \varepsilon(t, \chi\pi \times \chi'\pi') L(1/qt, \chi^{-1}\pi^\vee \times \chi'^{-1}\pi'^{-1})$$

and

$$L(t, \rho \otimes \rho' \otimes \chi\chi') = \varepsilon(t, \rho \otimes \rho' \otimes \chi\chi') L(1/qt, \rho^\vee \otimes \rho'^\vee \otimes \chi^{-1}\chi'^{-1}).$$

Proposition 12.14 then implies that at each place $v \in |C|$, we have

$$\frac{L(t, \pi_v \times \pi'_v)}{\varepsilon(t, \pi_v \times \pi'_v, \psi_v) L(1/qt, \pi_v^\vee \times \pi_v'^\vee)} = \frac{L(t, \rho_v \otimes \rho'_v)}{\varepsilon(t, \rho_v \otimes \rho'_v, \psi_v) L(1/qt, \rho_v^\vee \times \rho_v'^\vee)}.$$

By Prop. 12.25, the poles of $L(t, \rho_v \otimes \rho'_v)$ and $L(1/qt, \rho_v^\vee \times \rho_v'^\vee)$ occur at different places, and their absolute values are powers of $q^{1/2}$.

Let v be a place where π_v is ramified. Let τ be a cuspidal representation of a standard Levi subgroup $M(F_v)$ of $\mathrm{GL}(r, F_v)$ with central character of finite order, such that π_v is a constituent of the representation of $\mathrm{GL}(r, F_v)$ normalizedly induced from τ on the associated standard parabolic subgroup $P(F_v)$. By Lemma 12.20 there is a cuspidal representation π' of $M(\mathbb{A})$ with central character of finite order and cuspidal component at ∞, with $\pi'_v = \tau^\vee$. As π_v is unitarizable and generic, and π'_v is tempered, Corollary 12.11 asserts that $L(t, \pi_v \times \pi'_v)$ and $L(1/qt, \pi_v^\vee \times \pi_v'^\vee)$ do not share a pole. Hence

$$L(t, \pi_v \times \pi'_v) = L(t, \rho_v \otimes \rho'_v)$$

and the poles of $L(t, \pi_v \times \pi'_v) = L(t, \pi_v \times \tau^\vee)$ have absolute values powers of $q^{1/2}$. By Theorems 12.9 and 12.10 we conclude that π_v is tempered.

If $\pi' \in A^{r'}(F, \infty)$ and $r' \le r$, we have that π'_v is also tempered. By Corollary 12.11 $L(t, \pi_v \times \pi'_v)$ and $L(1/qt, \pi_v^\vee \times \pi_v'^\vee)$ do not share a pole. Hence for each $v \in |C|$ we obtain the equalities of L and ε factors asserted in the theorem. $\square$

12.7. **Local and Global Correspondence.** The local correspondence will now be deduced from Theorem 12.17 (or 11.1), which asserts the existence of the correspondence $\pi \mapsto \rho$. This local correspondence will be used to state in detail and prove the global correspondence of Theorem 12.18.

Let F_v be a local field of characteristic $p > 0$. Fix a prime $\ell \neq p$. Recall that we fix an isomorphism $\overline{\mathbb{Q}}_\ell \simeq \mathbb{C}$. Let $S_\ell^r(F_v)^+$ denote the set of isomorphism classes of ℓ-adic sheaves of rank r on $\operatorname{Spec} F_v$ whose determinant is of finite order. Let $S_\ell^r(F_v)$ be the subset of irreducible such sheaves. We use the notations $S_\ell^r(F_v)^+$ and $S_\ell^r(F_v)$ also for the sets of isomorphism classes of the representations σ of $\operatorname{Gal}(\overline{F}_v/F_v)$ associated (see [DF], 1.1(c)) with the sheaves ρ. Write $\sigma^{\text{F-ss}}$ for the Frob-semisimple ([De2], Section 8) representation attached to the representation σ.

Lemma 12.30.   (1) *Let ρ_v be a nonzero ℓ-adic sheaf on $\operatorname{Spec} F_v$. It is irreducible iff the local L-function $L(t, \rho_v \otimes \rho_v^\vee)$ has poles only on $|t| = 1$ and has a simple pole at $t = 1$.*

(2) *Let ρ_v and ρ_v' be two irreducible ℓ-adic sheaves on $\operatorname{Spec} F_v$. They are isomorphic iff $L(t, \rho_v' \otimes \rho_v^\vee)$ has a simple pole at $t = 1$.*

Proof. Let $\sigma^0(\operatorname{St}_r)$ be the unique [De2] Frob-semisimple indecomposable ℓ-adic representation of $\operatorname{Gal}(\overline{F}_v/F_v)$ such that $\operatorname{gr}_j^M \sigma^0(\operatorname{St}_r)$ is $\overline{\mathbb{Q}}_\ell(-(j+r-1)/2)$ if $j - r - 1$ is even and $|j| \leq r - 1$, and 0 otherwise. The Tate twist $\operatorname{Sp}_r(1) = \sigma^0(\operatorname{St}_r)((r-1)/2)$ is called the *special* representation of dimension r of $\operatorname{Gal}(\overline{F}_v/F_v)$. It is indecomposable and a successive extension of one-dimensional representations

$$\overline{\mathbb{Q}}_\ell((1-r)/2), \quad \overline{\mathbb{Q}}_\ell((1-r)/2 + 1), \quad \ldots, \overline{\mathbb{Q}}_\ell((r-1)/2),$$

with $\overline{\mathbb{Q}}_\ell((r-1)/2)$ as the unique irreducible submodule and with $\overline{\mathbb{Q}}_\ell((1-r)/2)$ as the unique irreducible quotient module.

The Frob-semisimplification $\sigma^{\text{F-ss}}$ of any ℓ-adic representation σ of dimension r of $\operatorname{Gal}(\overline{F}_v/F_v)$ whose determinant is of finite order can be written as a direct sum

$$\oplus_{n \geq 1} \oplus_{1 \leq r' \leq r} \oplus_{\tau' \in S_\ell^{r'}(F_v)} (\operatorname{Sp}_n(1) \otimes \tau')^{m(n, \tau')}$$

for some uniquely determined integers $m(n, \tau') \geq 0$ ([De1], (3.1.3)(ii)), which are almost all 0. Its L-function is

$$L(t, \sigma) = L(t, \sigma^{\text{F-ss}}) = \prod_{n \geq 1} \prod_{1 \leq r' \leq r} \prod_{\tau' \in S_\ell^{r'}(F_v)} L(t, \operatorname{Sp}_n(1) \otimes \tau')^{m(n, \tau')}$$

$$= \prod_{n \geq 1} \prod_{1 \leq r' \leq r} \prod_{\tau' \in S_\ell^{r'}(F_v)} L(tq^{(n-1)/2}, \tau')^{m(n, \tau')}.$$

If $\tau' \in S_\ell^{r'}(F_v)$ then $L(t, \tau')$ is 1 unless $r' = 1$ and τ' is an unramified character χ of finite order, in which case $L(t, \chi)$ has no zeroes, its poles are on the circle $|t| = 1$, and there is a pole at $t = 1$ iff $\chi = 1$. Applying the last displayed formula to $\sigma = \sigma_1 \otimes \sigma_2$ it follows that for $\sigma_i \in S_\ell^{r_i}(F_v)$, $i = 1$,

2, we have that $L(t, \sigma_1 \otimes \sigma_2)$ is not identically 1 iff $r_1 = r_2$ and there is an unramified character χ of finite order with $\sigma_2 \simeq \sigma_1^\vee \cdot \chi$.

By [De3], (1.6.11.2), we have

$$\mathrm{Sp}_a(1) \otimes \mathrm{Sp}_b(1) = \oplus_{0 \le j < \min(a,b)} \mathrm{Sp}_{a+b-1-2j}(1).$$

Hence for $\sigma_1, \sigma_2 \in S_\ell^r(F_v)$ with $\sigma_1^{\text{F-ss}}, \sigma_2^{\text{F-ss}}$ decomposing as above, we have

$$L(t, \sigma_1 \otimes \sigma_2) = \prod_{n_1, n_2 \ge 1} \prod_{1 \le r_1', r_2' \le r} \prod_{\tau_1' \in S_\ell^{r_1'}(F_v), , \tau_2' \in S_\ell^{r_2'}(F_v)} \prod_{0 \le j < \min(n_1, n_2)}$$
$$L(tq^{(n_1+n_2)/2-1-j}, \tau_1' \otimes \tau_2')^{m_1(n_1, \tau_1') m_2(n_2, \tau_2')}$$
$$= \prod_{n_1, n_2 \ge 1} \prod_{1 \le r' \le r} \prod_{\tau' \in S_\ell^{r'}(F_v)} \prod_\chi \prod_j L(tq^{(n_1+n_2)/2-1-j}, \chi)^{m_1(n_1, \tau') m_2(n_2, \tau'^\vee \chi)},$$

where χ ranges over the set of unramified characters of F_v of finite order. Since $\mathrm{Sp}_n(1)^\vee \simeq \mathrm{Sp}_n(1)$ ([De3], (1.6.11.3)), for $\sigma \in S_\ell^r(F_v)$ we obtain that $L(t, \sigma \otimes \sigma^\vee)$ equals

$$\prod_{n_1, n_2 \ge 1} \prod_{1 \le r' \le r} \prod_{\tau' \in S_\ell^{r'}(F_v)} \prod_\chi \prod_{0 \le j < \min(n_1, n_2)} L(tq^{(n_1+n_2)/2-1-j}, \chi)^{m_1(n_1, \tau') m_2(n_2, \tau'^\vee)}.$$

Write σ' for $\sigma^{\text{F-ss}}$. Then $\sigma' \otimes \sigma'^\vee$ is the Frob-semisimple representation associated with $\sigma \otimes \sigma^\vee$ and $L(t, \sigma \otimes \sigma^\vee)$ equals $L(t, \sigma' \otimes \sigma'^\vee)$. If this L-function has a simple pole at $t = 1$ then we deduce from the last displayed formula that σ' is irreducible. Hence σ is irreducible, as required. Indeed, if σ' is irreducible, then its restriction—and so also that of σ—to the inertia subgroup of $\mathrm{Gal}(\overline{F}_v/F_v)$, factors via a finite quotient. In this case σ' is the semisimplification of σ. Thus when σ' is irreducible, so is σ, and $\sigma' = \sigma$. Conversely, if σ is irreducible, then it factors via a finite quotient. Hence it is Frob-semisimple.   □

Denote by $A^r(F_v)$ the set of equivalence classes of cuspidal (by which we mean irreducible admissible such) representations of $\mathrm{GL}(r, F_v)$. Fix a nontrivial character $\psi_v : F_v \to \overline{\mathbb{Q}}_\ell^\times$. We have

Theorem 12.31. *There exists a unique series, indexed by $r \ge 1$, of bijections $A^r(F_v) \to S_\ell^r(F_v)$, $\pi_v \mapsto \rho(\pi_v)$, satisfying, for all π_v in $A^r(F_v)$, (1) $\rho(\pi_v^\vee) = \rho(\pi_v)^\vee$; (2) $\det \rho(\pi_v)$ corresponds to the central character ω_{π_v} of π_v by local class field theory; (3) for any character χ_v of $F_v^\times$ of finite order, $\rho(\pi_v \chi_v) = \rho(\pi_v) \chi_v$; and (4) for every $r' \ge 1$ and π_v' in $A^{r'}(F_v)$, we have $L(t, \rho(\pi_v) \otimes \rho(\pi_v')) = L(t, \pi_v \times \pi_v')$ and $\varepsilon(t, \rho(\pi_v) \otimes \rho(\pi_v'), \psi_v) = \varepsilon(t, \pi_v \times \pi_v', \psi_v)$.*

Proof. By induction on r, suppose $\pi_v \mapsto \rho(\pi_v)$, satisfying (1)–(3) and (4) for $r' \le r$, has been constructed in ranks $< r$, where $r \ge 2$. We define it for r. For that, we view F_v as the completion of a function field F of a smooth projective absolutely irreducible curve C, at a place $v \in |C|$. Given a cuspidal representation π_v of $\mathrm{GL}(r, F_v)$ with central character of finite order, there exists—by Lemma 12.20—a cuspidal representation $\pi \in A_\ell^r(F, \infty)$ of

$GL(r, \mathbb{A})$, with central character of finite order, whose component at v is our π_v. By Theorem 12.17 our π corresponds to an ℓ-adic sheaf $\rho(\pi) \in S_\ell^r(F, \infty)$. By Theorem 12.29 $L(t, \rho(\pi)_v \otimes \rho(\pi)_v^\vee) = L(t, \pi_v \times \pi_v^\vee)$. By Lemma 12.30 (1), $\rho(\pi)_v$ is irreducible. If $\pi' \in A^r(F, \infty)$ is another cuspidal representation whose component at v is $\pi_v' = \pi_v$, then $L(t, \rho(\pi')_v \otimes \rho(\pi)_v^\vee) = L(t, \pi_v' \times \pi_v^\vee)$. By Lemma 12.30 (2), $\rho(\pi')_v$ is equivalent to $\rho(\pi)_v$. We thus defined a map $\pi_v \mapsto \rho(\pi_v) = \rho(\pi)_v$ of $A^r(F_v)$ into $S_\ell^r(F_v)$.

Given $\pi_v \in A^r(F_v)$ and $\pi_v' \in A^{r'}(F_v)$ with $r' \leq r$, we can view them as components of $\pi \in A^r(F, \infty)$ and $\pi' \in A^{r'}(F, \infty)$. By Theorem 12.29 we have

$$L(t, \rho(\pi_v) \otimes \rho(\pi_v')) = L(t, \pi_v \times \pi_v'), \qquad \varepsilon(t, \rho(\pi_v) \otimes \rho(\pi_v', \psi_v) = \varepsilon(t, \pi_v \times \pi_v', \psi_v).$$

In particular we have $L(t, \rho(\pi_v') \otimes \rho(\pi_v)^\vee) = L(t, \pi_v' \times \pi_v^\vee)$. So Lemma 12.30 (2) implies that the map $A^r(F_v) \to S_\ell^r(F_v)$, $\pi_v \mapsto \rho(\pi_v)$, is injective. Clearly it satisfies (1)–(3).

The map $\pi_v \mapsto \rho(\pi_v)$ is bijective, due to a counting argument recorded as Theorem 15.17 in [LRS]. The map with these properties (1)–(4) is unique by [H]. $\qquad\qquad\square$

We now repeat Theorem 12.18.

Theorem 12.32. *For every positive integer r, the correspondence defines a bijection $A^r(F, \infty) \to S_\ell^r(F, \infty)$, $\pi \mapsto \rho_\pi$. Moreover, $\pi \in A^r(F, \infty)$ and the corresponding $\rho \in S_\ell^r(F, \infty)$ are ramified at the same places.*

Proof. The case of $r = 1$ is class field theory, so we assume $r \geq 2$. In particular we can identify characters of $\Gamma^\times \backslash \mathbb{A}^\times$ of finite order with ℓ-adic sheaves of rank 1 and finite order. This identification respects L and ε factors at all places. By Chebotarev density theorem, to π corresponds at most one ρ. By rigidity theorem for $GL(r)$ there is at most one cuspidal π corresponding to a given ρ. We may consider only π whose central character is of finite order and ρ whose determinant is of finite order (thus $(\det \rho)^{\otimes m} \simeq \mathbb{Q}_\ell$ for some integer $m > 0$).

The map $A^r(F, \infty) \to S_\ell^r(F, \infty)$, $\pi \mapsto \rho_\pi$, has already been constructed using the moduli scheme of elliptic modules, its étale ℓ-adic cohomology with compact support and coefficients in a sheaf, Deligne's conjecture, and the trace formula. In particular we have the Ramanujan conjecture too: each unramified component of $\pi \in A^r(F, \infty)$ is tempered. What remains to be seen is that the map $\pi \mapsto \rho_\pi$ is surjective.

Let $\rho \in S_\ell^r(F, \infty)$ be an irreducible ℓ-adic sheaf. Denote by S the finite set of places $v \in |C|$, $v \neq \infty$, where ρ is ramified. At each $v \notin S$, $v \neq \infty$, denote by $\mathbf{z}(\rho_v)$ the r-tuple of its Frobenius eigenvalues $(z_1(\rho_v), \ldots, z_r(\rho_v))$, where the $z_i(\rho_v)$ are viewed as complex numbers ordered to satisfy $|z_i(\rho_v)| \geq |z_{i+1}(\rho_v)|$ $(1 \leq i < r)$. The corresponding G_v-module $\pi_v = I(\mathbf{z}(\rho_v))$, normalizedly induced from the unramified character (of the upper triangular Borel subgroup) defined by $\mathbf{z}(\rho_v)$, is unramified, and possibly reducible (if $z_i(\rho_v) = q_v z_{i+1}(\rho_v)$ for some i $(1 \leq i < r)$).

Lemma 12.33. *The G_v-module $\pi_v = I(\mathbf{z}(\rho_v))$ is nondegenerate.*

Proof. Let w_0 be a nonzero $GL(r, R_v)$-fixed vector in the space of π_v (w_0 is unique up to a scalar). Let W_0 denote the Whittaker function associated to π_v and ψ_v (which is taken to be unramified) in [Sh]. Putting $F(\pi_v(g)w_0) = W_0(g)$, we obtain a nonzero morphism F of G_v-modules from π_v to the space $\mathrm{Ind}(\psi_v; N_v, G_v)$ of Whittaker functions. To show that π_v is nondegenerate, we need to show that F is injective. But this follows from a result of [Z], which asserts that π_v has a unique irreducible nondegenerate subquotient π_v^0, which is in fact the unique subrepresentation of π_v. Namely the irreducible nondegenerate subrepresentation π_v^0 of π_v is a subrepresentation of any subrepresentation of π_v. Now if $F : \pi_v \to \mathrm{Ind}(\psi_v; N_v, G_v)$ had a nontrivial kernel K, it would contain π_v^0, and there results an embedding of the degenerate G_v-module π_v/K in the space $\mathrm{Ind}(\psi_v; N_v, G_v)$ of Whittaker functions. This is a contradiction to the assumption that F is not injective, and the lemma follows.   □

Remark 12.7. Had we arranged the eigenvalues so that $|z_i(\rho_v)| \leq |z_{i+1}(\rho_v)|$, then π_v would have the nondegenerate constituent π_v^0 as a unique quotient. In this case the embedding of π_v^0 in $W_v = \mathrm{Ind}(\psi_v; N_v, G_v)$ extends to a morphism of π_v into W_v whose kernel is the subrepresentation K_v of π_v with $\pi_v/K_v \cong \pi_v^0$. In particular the resulting morphism $\pi_v \to W_v$ is not injective if π_v is reducible.

Now given $\rho \in S_\ell^r(F, \infty)$, we defined an unramified generic representation π_v of $GL(r, F_v)$ for each $v \in |C|$, $v \neq \infty$, $v \notin S$, where ρ is unramified. At ∞, let π_∞ be the cuspidal representation of $GL(r, F_\infty)$ associated with the restriction ρ_∞ to $\mathrm{Spec}\, F_\infty$ by the local correspondence. At each place $v \in S$ choose a generic irreducible representation π_v of $GL(r, F_v)$ whose central character ω_{π_v} corresponds to $\det(\rho_v)$. Then $\pi = \otimes_v \pi_v$ ($v \in |C|$) is an admissible irreducible generic representation of $GL(r, \mathbb{A})$ whose central character $\omega_\pi = \otimes_v \omega_{\pi_v}$ corresponds to $\det \rho$ and its component π_∞ is cuspidal.

Let $\chi = \otimes_v \chi_v$ ($v \in |C|$) be a character of $\mathbb{A}^\times/F^\times$ of finite order which is highly ramified at the places $v \in S$. We aim to show for any cuspidal representation $\pi' = \otimes_{v \in |C|} \pi_v'$ in $A^{r'}(F, \infty)$, $r' < r$, such that π_v' ($v \in S$) are unramified, that the formal power series $L(t, \chi\pi \times \pi')$ and $L(t, \chi^{-1}\pi^\vee \times \pi'^\vee)$ are polynomials satisfying the functional equation

$$L(t, \chi\pi \times \pi') = \varepsilon(t, \chi\pi \times \pi')L(t, \chi^{-1}\pi^\vee \times \pi'^\vee).$$

Theorem 13.1 would then apply to imply the existence of a cuspidal $\pi \in A^r(F, \infty)$ corresponding to ρ, proving the theorem.

By induction the theorem holds for r'. Thus π' corresponds to an ℓ-adic sheaf $\rho' = \rho_{\pi'} \in S_\ell^{r'}(F, \infty)$, and $\pi'\chi$ corresponds to $\rho' \otimes \chi$, and their L and ε factors are equal at all places.

At the places $v \notin S \cup \{\infty\}$, the factor π_v is unramified. Hence by definition of π_v, we have

$$L(t, \chi_v\pi_v \times \pi_v') = L(t, \rho_v \otimes \chi_v \otimes \rho_v'), \qquad L(t, \chi_v^{-1}\pi_v^\vee \times \pi_v'^\vee) = L(t, \rho_v^\vee \chi_v^{-1} \otimes \rho_v'^\vee),$$

$$\varepsilon(t, \chi_v\pi_v \times \pi_v', \psi_v) = \varepsilon(t, \rho_v \otimes \chi_v \otimes \rho_v', \psi_v).$$

At the places $v \in S$ the factor π'_v is unramified. Choosing χ_v sufficiently ramified with respect to π_v and ρ_v, we have

$$L(t, \chi_v \pi_v \times \pi'_v) = 1 = L(t, \rho_v \otimes \chi_v \otimes \rho'_v), \ L(t, \chi_v^{-1} \pi_v^\vee \times \pi_v'^\vee) = 1 = L(t, \rho_v^\vee \chi_v^{-1} \otimes \rho_v'^\vee),$$

$$\varepsilon(t, \chi_v \pi_v \times \pi'_v, \psi_v) = \varepsilon(t, \chi_v, \psi_v)^{rr'-1} \varepsilon(t, \chi_v \omega_{\pi_v}^{r'} \omega_{\pi'_v}^r, \psi_v)$$

$$= \varepsilon(t, \chi_v, \psi_v)^{rr'-1} \varepsilon(t, \chi_v \det(\rho_v)^{r'} \det(\rho'_v)^r, \psi_v) = \varepsilon(t, \rho_v \otimes \chi_v \otimes \rho'_v, \psi_v).$$

The same holds at ∞ as π_∞ and π'_∞ are cuspidal, by the local correspondence and the choice of ρ_∞ and ρ'_∞.

We deduce that $L(t, \chi\pi \times \pi') = L(t, \rho \otimes \chi \otimes \rho')$,

$$L(t, \chi^{-1} \pi^\vee \times \pi'^\vee) = L(t, \rho^\vee \otimes \chi^{-1} \otimes \rho'^\vee),$$

and using the product formula for the local factors that

$$\varepsilon(t, \chi\pi \times \pi') = \varepsilon(t, \rho \otimes \chi \otimes \rho').$$

Transferring from ρ, ρ' to π, π' we see that $L(t, \chi\pi \times \pi')$ and $L(t, \rho \otimes \chi \otimes \rho')$ are rational functions which satisfy

$$L(t, \chi\pi \times \pi') = \varepsilon(t, \chi\pi \times \pi') L(1/qt, \chi^{-1}\pi^\vee \times \pi'^\vee).$$

It remains to see that they are polynomials. This follows from Grothendieck's cohomological interpretation of the L-function:

$$L(t, \rho \otimes \chi \otimes \rho') = \prod_{0 \le i \le 2} \det[I - t \cdot \mathrm{Fr} \,| H_c^i(\rho \otimes \chi \otimes \rho')]^{(-1)^{i+1}},$$

$$L(t, \rho^\vee \otimes \chi^{-1} \otimes \rho'^\vee) = \prod_{0 \le i \le 2} \det[I - t \cdot \mathrm{Fr} \,| H_c^i(\rho^\vee \otimes \chi^{-1} \otimes \rho'^\vee)]^{(-1)^{i+1}}.$$

Indeed, since ρ and ρ' are irreducible of ranks $r \ne r'$, the $H_c^i(\rho \otimes \chi \otimes \rho')$ and $H_c^i(\rho^\vee \otimes \chi^{-1} \otimes \rho'^\vee)$ vanish for $i = 0, 2$. Theorem 13.1 now implies that there is an automorphic representation whose components outside S are the same as those of $\chi\pi$. Hence up to changing the factors of π at $v \in S$, this π is automorphic, corresponding to ρ. Its component at ∞ is cuspidal, so π is cuspidal. $\qquad\square$

Let us verify that the local and global correspondences are compatible.

Proposition 12.34. *Let $\rho \in S_\ell^r(F, \infty)$ be an irreducible ℓ-adic sheaf on $\mathrm{Spec}\, F$ of rank $r \ge 2$ whose restriction to $\mathrm{Spec}\, F_\infty$ is irreducible. Let $\pi \in A^r(F, \infty)$ be the corresponding cuspidal representation of $\mathrm{GL}(r, \mathbb{A})$. Its component π_∞ at ∞ is cuspidal. Then the local factor π_v of π at $v \in |C|$ is the unique generic irreducible admissible representation of $\mathrm{GL}(r, F_v)$ whose central character ω_{π_v} corresponds to $\det \rho_v$ by local class field theory, and such that for any integer $r' < r$ and any $\pi' \in A^{r'}(F, \infty)$, denoting by $\rho' \in S_\ell^{r'}(F, \infty)$ the corresponding ℓ-adic sheaf, we have*

$$L(t, \pi_v \times \pi'_v) = 1 = L(t, \rho_v \otimes \rho'_v), \quad L(t, \pi_v^\vee \times \pi_v'^\vee) = L(t, \rho_v^\vee \otimes \rho_v'^\vee),$$

$$\varepsilon(t, \pi_v \times \pi'_v, \psi_v) = \varepsilon(t, \rho_v \otimes \rho'_v, \psi_v).$$

Proof. By Theorem 12.29 we already know that the component π_v of π at v satisfies these properties. We need to show the uniqueness of π_v. Thus suppose the irreducible admissible generic representation π_v'' of $\mathrm{GL}(r, F_v)$ satisfies these properties. Denote by π'' the irreducible admissible representation of $\mathrm{GL}(r, \mathbb{A})$ whose local component at v is π_v'' and its other components are the same as those of π.

The representation π'' has the same central character as π. The Euler product which defines its L-function is the same as that of π. Hence it converges absolutely at some disc, and for every cuspidal representation $\pi' \in A^{r'}(F, \infty)$ of $\mathrm{GL}(r', \mathbb{A})$ we have $L(t, \pi'' \times \pi') = L(t, \pi \times \pi')$,

$$L(t, \pi''^\vee \times \pi'^\vee) = L(t, \pi^\vee \times \pi'^\vee), \qquad \varepsilon(t, \pi'' \times \pi') = \varepsilon(t, \pi \times \pi').$$

Hence $L(t, \pi'' \times \pi')$ and $L(t, \pi''^\vee \times \pi'^\vee)$ are polynomials which satisfy the functional equation

$$L(t, \pi'' \times \pi') = \varepsilon(t, \pi'' \times \pi') L(1/qt, \pi''^\vee \times \pi'^\vee).$$

By the simple converse Theorem 13.1 we see that π'' is a cuspidal representation of $\mathrm{GL}(r, \mathbb{A})$. It coincides with π at all places except possibly at v. Hence by the rigidity theorem for cuspidal representations of $\mathrm{GL}(r, \mathbb{A})$ we deduce that $\pi_v'' = \pi_v$, and the required uniqueness follows.   □

Using the similarity of the construction of smooth ℓ-adic sheaves of rank r on $\mathrm{Spec}\, F_v$ and the classification of admissible representations of $\mathrm{GL}(r, F_v)$ by [BZ], one can extend the local correspondence from the case of irreducible and cuspidal case to that of maps, indexed by $r \geq 1$, from the set of isomorphism classes of ℓ-adic sheaves of rank r on $\mathrm{Spec}\, F_v$ (whose determinant is of finite order) to the set of equivalence classes of irreducible admissible representations of $\mathrm{GL}(r, F_v)$ (whose central character is of finite order). These maps preserve the local L and ε factors of pairs and are compatible with taking contragredient, twisting with characters of finite order, and local class field theory. These maps are surjective and two ℓ-adic sheaves of the same rank on $\mathrm{Spec}\, F_v$ have the same image iff they have the same F-semisimplifications ([De2], Section 8).

From the last proposition we then conclude

Corollary 12.35. *Let $\rho \in S_\ell^r(F, \infty)$ be an irreducible ℓ-adic sheaf on $\mathrm{Spec}\, F$ of rank $r \geq 2$ whose restriction to $\mathrm{Spec}\, F_\infty$ is irreducible. Let $\pi \in A^r(F, \infty)$ be the corresponding cuspidal representation of $\mathrm{GL}(r, \mathbb{A})$. Then the component π_v of π at v is the image of ρ_v under the local correspondence for F_v.*

We could have approached the construction of the π corresponding to ρ differently, namely define π to be $\otimes_v \pi_v$ where each π_v is defined by the local correspondence from ρ_v, also when ρ_v is not unramified or irreducible. For this we would need to expand the paragraph preceding the last corollary.

Let ∞ be a fixed place of F.

Corollary 12.36. *Let ρ be a λ-adic irreducible constructible representation of $W(\overline{F}/F)$ with determinant of finite order. Suppose that the restriction $\rho_\infty = \rho | W(\overline{F}_\infty/F_\infty)$ is irreducible. Then for all v where ρ_v is unramified,* *the roots of $P_{v,\rho}(t)$ (namely the eigenvalues of the Frobenius) have complex absolute value one.*

Proof. The $\mathrm{GL}(r,\mathbb{A})$-module π corresponding to ρ is cuspidal with a cuspidal component at ∞. Hence π satisfies the purity Theorem 10.8, namely the Hecke eigenvalues of π_v are units, as required. $\qquad\square$

Let ℓ be a rational prime and $\rho : W(\overline{F}/F) \to \mathrm{GL}(r,\overline{\mathbb{Q}}_\ell)$ an irreducible ℓ-adic representation of the Weil group of F. As noted in Lemma 12.1, there is a finite extension E_λ of $\mathbb{Q}_\ell$ such that ρ factorizes through $\mathrm{GL}(r,E_\lambda)$. Replacing ρ by $\rho \otimes \chi$ for some nowhere ramified character χ of $W(\overline{F}/F)$, we may assume that $\det \rho$ has finite order and consequently that ρ extends to an ℓ-adic representation $\rho : \mathrm{Gal}(\overline{F}/F) \to \mathrm{GL}(r,E_\lambda)$ of the Galois group. Fix a place ∞ of F.

Corollary 12.37. *Let $\rho : \mathrm{Gal}(\overline{F}/F) \to \mathrm{GL}(r,E_{\lambda'})$ be an irreducible constructible λ'-adic representation of F, whose restriction to $\mathrm{Gal}(\overline{F}_\infty/F_\infty)$ is irreducible, with determinant of finite order. Then there exists a finite extension $\mathbb{Q}(\rho)$ of $\mathbb{Q}$ such that each of the eigenvalues of $\rho_v(\mathrm{Fr}_v)$ lies in $\mathbb{Q}(\rho)$ for every place v of F such that $\rho_v = \rho| \mathrm{Gal}(\overline{F}_v/F_v)$ is unramified. Moreover, there exists a finite set $V(\rho)$ of rational primes excluding the residual characteristic ℓ' of $E_{\lambda'}$ and a compatible (see [De2] or [S2]) family $\{\rho_\ell : \mathrm{Gal}(\overline{F}/F) \to \mathrm{GL}(r,\mathbb{Q}(\rho)_\ell); \ \ell \notin V(\rho)\}$, where $\mathbb{Q}(\pi)_\ell$ is any completion of $\mathbb{Q}(\pi)$ over $\mathbb{Q}_\ell$, with $\rho = \rho_{\ell'}$ (in particular $E_{\lambda'}$ is an extension of $\mathbb{Q}(\pi)_{\ell'}$) and such that $\rho_{\ell,\infty}$ is irreducible for all ℓ.*

Proof. Let π be the cuspidal $G(\mathbb{A})$-module which corresponds to ρ. Put $\mathbb{Q}(\rho) = \mathbb{Q}(\pi)$, where $\mathbb{Q}(\pi)$ is the field of definition of π which is introduced at the end of Chap. 10. Then $\mathbb{Q}(\pi_v) = \mathbb{Q}(\mathrm{tr}\,\rho_v(\mathrm{Fr}_v^m); \ m \leq 1)$ lies in $\mathbb{Q}(\rho)$ for all v where ρ_v is unramified. The field $\mathbb{Q}(\pi)$ is a finite extension of $\mathbb{Q}$ since $\mathbb{Q}(\det \rho)$ is a finite extension of $\mathbb{Q}$ by assumption. This proves the first assertion; the second follows at once from Theorem 12.17. $\qquad\square$

13. SIMPLE CONVERSE THEOREM

We prove a simple form of the converse theorem for $GL(n)$ over a function field F, "simple" referring to a cuspidal component. Thus a generic admissible irreducible representation π of the adèle group $GL(n, \mathbb{A})$ with cuspidal components at a finite nonempty set S of places of F whose product L-function $L(t, \pi \times \pi')$ is a polynomial in t and has a functional equation for each cuspidal representation π' of $GL(n - 1, \mathbb{A})$ whose components at S are cuspidal, is automorphic, necessarily cuspidal. The usual form of the converse theorem deals with the case where S is empty. But our simple form is sufficient for applications of the simple trace formula.

13.1. Introduction.

Theorem 13.1. *Let X be a smooth projective absolutely irreducible curve over a finite field $\mathbb{F}_q$. Let F be its function field, $\mathbb{A}$ the ring of adèles, $n > 2$ an integer, and $S_1 \neq \emptyset$ and S_2 disjoint finite sets of places of F. Let $\pi = \otimes_v \pi_v$ (v ranges over the set $|X|$ of places of F) be a unitarizable irreducible admissible (in particular π_v is unramified for almost all v) locally generic (thus π_v is generic for all v) representation of $GL(n, \mathbb{A})$ whose central character is trivial on $F^\times$ and whose components π_v, $v \in S_1$, are cuspidal. Suppose that for each cuspidal (by which we mean in particular irreducible and automorphic) representation π' of $GL(n - 1, \mathbb{A})$ whose component π'_v at $v \in S_1$ is cuspidal and at $v \in S_2$ is unramified, the formal series*

$$(13.1) \qquad L(t, \pi \times \pi') \quad and \quad L(t, \pi^\vee \times \pi'^\vee)$$

are polynomials satisfying the functional equation

$$(13.2) \qquad L(t, \pi \times \pi') = \varepsilon(t, \pi \times \pi') L(1/qt, \pi^\vee \times \pi'^\vee).$$

Then there exists a cuspidal representation π' of $GL(n, \mathbb{A})$ with $\pi'_v \simeq \pi_v$ for all $v \notin S_2$.

Extending work of Weil and others, Piatetski-Shapiro discussed in an unpublished manuscript of 1976 a variant of the statement above, named the converse theorem since when π is cuspidal, $L(t, \pi \times \pi')$ satisfies the functional equation. This converse theorem can be used to prove by induction automorphy of a product $\otimes_v \pi_v$ of representations π_v of $GL(n, F_v)$. Sometimes the induction assumption is satisfied only for representations π' which are cuspidal at the places $v \in S_1$. Such a situation has acquired the label "simple", as in such a case the trace formula simplifies considerably; see, e.g., [FK2]. Thus we name Theorem 13.1 a simple converse theorem. We discussed it – following Piatetski-Shapiro's exposition of 1976 – in the unpublished manuscript of 1983 which dealt mainly with applications underlying the present work. An extension to include the number field case appeared in [CPS] of 1994. The function field case of that appeared in an appendix to [Lf2] of 2002. It was used in [Lf2] to prove the reciprocity law between irreducible n-dimensional representations of the Galois group of F and cuspidal representations of $GL(n, \mathbb{A})$. However, a

Y.Z. Flicker, *Drinfeld Moduli Schemes and Automorphic Forms: The Theory of Elliptic Modules with Applications*, SpringerBriefs in Mathematics, DOI 10.1007/978-1-4614-5888-3_13, © Yuval Z. Flicker 2013

treatment of the simple converse theorem for $\mathrm{GL}(n)$ has not yet appeared. It is needed to obtain by relatively simple means a large part of the reciprocity law for $\mathrm{GL}(n)$ over function fields, a worthy aim in view of the intense technical difficulty of [Lf2]. This chapter is then written to address this lacuna. The number field case follows by combining our arguments with those of [CPS]. But we currently know of applications only in the function field case.

13.2. **Generic Representations.** Let X be a projective smooth absolutely irreducible curve over a finite field $\mathbb{F}_q$ of cardinality q and characteristic p. Let $F = \mathbb{F}_q(X)$ denote the function field of X over $\mathbb{F}_q$. The set $|X|$ of closed points of X is naturally isomorphic to the set of places v (isomorphism classes of absolute values $|.|_v$) of F. For each v in $|X|$ denote by F_v the completion of F by $|.|_v$, by R_v its ring of integers, by π_v a generator of the maximal ideal in the local ring R_v. Let q_v be the cardinality of the residue field $k_v = R_v/(\pi_v)$. Normalize the absolute value by $|\pi_v|_v = q_v^{-1}$. Define the valuation $\deg_v : F_v^\times \to \mathbb{Z}$ by $|x|_v = q_v^{-\deg_v(x)}$. Thus $\deg_v(\pi_v) = 1$.

The ring $\mathbb{A}$ of adèles of F is the restricted product $\prod_v F_v$ of the F_v ($v \in |X|$) with respect to the compact subrings R_v. If S is a finite subset of $|X|$ put $F_S = \prod_{v \in S} F_v$. Then $\mathbb{A} = \bigcup_S F_S \cdot \prod_{v \notin S} R_v$. Put $\mathbb{A}^S = \prod_{v \notin S} F_v$. Then $\mathbb{A} = F_S \cdot \mathbb{A}^S$. The group of idèles of F is $\mathbb{A}^\times = \prod_v F_v^\times$, the restricted product of the multiplicative groups $F_v^\times$ of F_v with respect to the compact groups $R_v^\times$. Thus $\mathbb{A}^\times = \bigcup_S F_S^\times \cdot \prod_{v \notin S} R_v^\times$. It is the multiplicative group of $\mathbb{A}$, and $\mathbb{A}^\times = F_S^\times \cdot \mathbb{A}^{S\times}$.

Let B denote the upper triangular subgroup of $G = \mathrm{GL}(n)$, N its unipotent radical, A the diagonal subgroup, P' the standard (containing B) parabolic subgroup of type $(n-1, 1)$, P its (mirabolic) subgroup of matrices with bottom row $(0, \ldots, 0, 1)$, and $\overline{P} = {}^tP \subset \overline{P}' = {}^tP'$ the opposite mirabolic and parabolic subgroups with last column ${}^t(0, \ldots, 0, 1)$ and ${}^t(0, \ldots, 0, *)$. The center of G is denoted by Z. The index n is used to emphasize if needed. Thus we view G_{n-1} as the subgroup $P_n \cap \overline{P}_n$ of G_n, and consequently, e.g., $N_{n-1} \subset P_{n-1} \subset G_{n-1} \subset P_n \subset G_n$.

Let $\psi_n : F_v \to \mathbb{C}^\times$ ($v \in |X|$) be a character $\neq 1$. It defines a character, denoted again by ψ_n, of $N(F_v)$, by $\psi_v((u_{i,j})) = \psi_v(\sum_{1 \leq i < n} u_{i,i+1})$. We recall some local definitions and results from [GK] (see also [BZ]). Thus we omit the index v.

An admissible representation (π, V) of $G = \mathrm{GL}(n, F)$ (over $\mathbb{C}$) is called *generic* if there exists a nonzero linear form ℓ on V satisfying $\ell(\pi(u)\xi) = \psi(u)\ell(\xi)$ for all $u \in N(F)$ and $\xi \in V$. By [GK], Theorem C, if π is irreducible and ℓ exists, then ℓ is unique up to a scalar. By Theorem D, the space $W(\pi, \psi)$ of $W_\xi(g) = \ell(\pi(g)\xi)$, $\xi \in \pi$, makes under right translation a G-submodule of the induced representation $\mathrm{Ind}(G, N, \psi)$, such that $\pi \to W(\pi, \psi)$, $\xi \mapsto W_\xi$ is an isomorphism of G-modules. Theorem B asserts that every cuspidal representation of G is generic (by cuspidal representation we mean an irreducible one). By Schur's lemma ([BZ]), an irreducible π has a central character. Denote it by ω. Then in fact $W(\pi, \psi) \subset \mathrm{Ind}(G, ZN, \omega\psi)$, where $\omega\psi$ is the natural character on ZN.

Consider the restriction map $W(\pi, \psi) \hookrightarrow \mathrm{Ind}(P, N, \psi)$, $W \mapsto W|P$. Theorem E of [GK] asserts that if π is cuspidal then this map factorizes via $\mathrm{ind}(P, N, \psi) \hookrightarrow \mathrm{Ind}(P, N, \psi)$, and the deduced map $W(\pi, \psi) \hookrightarrow \mathrm{ind}(P, N, \psi)$ is an isomorphism of P-modules. Here ind indicates compact induction. Thus $\mathrm{ind}(P, N, \psi)$ consists of functions $f : P \to \mathbb{C}$ such that there exists an open subgroup $U_f \subset P$ with $f(upk') = \psi(u)f(p)$ ($u \in N$, $k' \in U_f$) and a compact subset S_f of $N\backslash P$ with $f(p) \neq 0$ ($p \in P$) implying $Np \in S_f$. Theorem F implies that if π is not cuspidal then $\{W|P; W \in W(\pi, \psi)\}$ strictly contains $\mathrm{ind}(P, N, \psi)$.

Lemma 13.2. *Let π be an irreducible generic representation of G. Denote by ω the central character of π. Then $W(\pi, \psi)$ ($\subset \mathrm{Ind}(G, ZN, \omega\psi)$) is contained in $\mathrm{ind}(G, ZN, \omega\psi)$ if and only if π is cuspidal.*

Proof. We use the decomposition $G = ZPK$, where $K = \mathrm{GL}(n, R)$. Then $\mathrm{supp}\, W \subset \mathrm{supp}(W|P) \cdot ZK$. If π is cuspidal then $\mathrm{supp}(W|P)$ is compact in $P \bmod N$ for any W in $W(\pi, \psi)$. Hence $\mathrm{supp}\, W$ is compact in $G \bmod ZN$. Conversely, if π is not cuspidal then the image of $W(\pi, \psi)$ under $W \mapsto W|P$ properly contains $\mathrm{ind}(P, N, \psi)$. Hence $W(\pi, \psi)$ is not contained in $\mathrm{ind}(G, ZN, \omega\psi)$. $\square$

Lemma 13.3. *Let π_n be an irreducible representation of $G_n = \mathrm{GL}(n, F)$. The image of the restriction map*

$$W \mapsto W|G_{n-1}, \qquad W(\pi_n, \psi_n) \to \mathrm{Ind}(G_{n-1}, N_{n-1}, \psi_{n-1})$$

contains $\mathrm{ind}(G_{n-1}, N_{n-1}, \psi_{n-1})$. The image is $\mathrm{ind}(G_{n-1}, N_{n-1}, \psi_{n-1})$ iff π_n is cuspidal.

Proof. The restriction map

$$W \mapsto W|P_n, \qquad W(\pi_n, \psi_n) \hookrightarrow \mathrm{Ind}(P_n, N_n, \psi_n)$$

has image $\mathrm{ind}(P_n, N_n, \psi_n)$ if π_n is cuspidal. The image strictly contains ind otherwise. The restriction map

$$W \mapsto W|G_{n-1}, \mathrm{ind}(P_n, N_n, \psi_n) \to \mathrm{ind}(G_{n-1}, N_{n-1}, \psi_{n-1}),$$

is an isomorphism of G_{n-1}-modules, and so is the map with ind replaced by Ind. $\square$

Let ω be a character of $F^\times$, hence also of $Z_{n-1} \simeq F^\times$. The map

$$A_\omega : \mathrm{ind}(G_{n-1}, N_{n-1}, \psi_{n-1}) \to \mathrm{ind}(G_{n-1}, Z_{n-1}N_{n-1}, \omega\psi_{n-1}),$$

$\phi \mapsto \overline{\phi}$, $\overline{\phi}(g) = \int_{Z_{n-1}} \phi(zg)\omega(z)^{-1}dz$, is onto. The same applies with ind replaced by Ind. We conclude:

Lemma 13.4. *Let π_n be an irreducible representation of G_n. The image of the map*

$$W \mapsto A_\omega(W|G_{n-1}), \qquad W(\pi_n, \psi_n) \to \mathrm{Ind}(G_{n-1}, Z_{n-1}N_{n-1}, \omega\psi_{n-1})$$

contains $\mathrm{ind}(G_{n-1}, Z_{n-1}N_{n-1}, \omega\psi_{n-1})$, with equality iff π_n is cuspidal. In particular, when π_n is cuspidal, the image of $W(\pi_n, \psi_n)$ under $W \mapsto A_\omega(W|G_{n-1})$ is cuspidal.

13.3. **The Global Functions** U **and** V. For the global theory we fix a character $\psi \neq 1$ of $\mathbb{A} \bmod F$. It defines a character, denoted again by ψ, of $N(\mathbb{A})/N(F)$, as in the local case, and for each $v \in |X|$, a restriction ψ_v to F_v and $N(F_v)$. We shall often write G for $G(F)$ and G_v for $G(F_v)$ (and for other algebraic groups). An admissible irreducible representation π of $G(\mathbb{A})$ (over $\mathbb{C}$) is the restricted product $\otimes_v \pi_v$ of irreducible admissible representations π_v of $G_v = G(F_v)$ which are almost all unramified (π_v is called *unramified* if its space contains a nonzero $K_v = G(R_v)$-fixed vector, which is necessarily unique up to a scalar). Namely the space of π is spanned by $\otimes_v \xi_v$, $\xi_v \in \pi_v$ for all v and ξ_v is a fixed K_v-fixed vector $\xi_v^0 \neq 0$ for almost all v. If π_v is irreducible, unramified, and generic, the vector ξ_v^0 can be chosen so that $W_v^0 = W_{\xi_v^0} \in W(\pi_v, \psi_v)$ satisfies $W_v^0(e) = 1$. The *Whittaker model* of a locally generic irreducible admissible π is $W(\pi, \psi) \subset \mathrm{Ind}(G(\mathbb{A}), N(\mathbb{A}), \psi)$, the space spanned by $W = \otimes_v W_v$ (which takes $g = (g_v)$ to $\prod_v W_v(g_v)$), where $W_v \in W(\pi_v, \psi_v)$ for all v and $W_v = W_v^0$ for almost all v. Then each W is smooth (right invariant under an open subgroup of $G(\mathbb{A})$) and satisfies

$$W(ug) = \psi(u)W(g) \quad (g \in G(\mathbb{A}), u \in N \backslash N(\mathbb{A})).$$

If $\omega = \otimes_v \omega_v$ is the central character of π then $W(zg) = \omega(z)W(g)$ $(z \in Z(\mathbb{A}))$.

Lemma 13.5. (1) *Let W be a Whittaker function on $G(\mathbb{A})$ (right smooth with $W(ug) = \psi(u)W(g)$ $(u \in N(\mathbb{A}), g \in G(\mathbb{A}))$. Then there exists a sequence $(m_v \in \mathbb{Z}; v \in |X|)$ with $m_v = 0$ for almost all v such that if $W(g) \neq 0$ for $g = bk$ with $b = (b_{i,j})$ in $B(\mathbb{A})$ and k in $K = \prod_v K_v$ then for all v and i $(1 \leq i < n)$ we have $|b_{i,i}/b_{i+1,i+1}|_v \leq q^{m_v}$.*
(2) *The sum*

$$U(g) = \sum_{p \in N \backslash P} W(pg) = \sum_{p \in N_{n-1} \backslash G_{n-1}} W(pg)$$

converges absolutely and uniformly on compact sets in $G(\mathbb{A})$.

Proof. (1) If not then there are v and i and big enough m_v such that $|b_{i,i}/b_{i+1,i+1}|_v > q^{m_v}$, and there is $u_v \in N(F_v)$ with $\psi_v(u_v) \neq 1$ such that $g_v^{-1}u_v g_v$ lies in an open subgroup of finite index in K_v under which W_v is right invariant. Then $\psi_v(u_v)W_v(g_v) = W_v(u_v g_v) = W_v(g_v(g_v^{-1}u_v g_v)) = W_v(g_v)$ implies $W_v(g_v) = 0$ and $W(g) = 0$.

(2) It follows from (1) that if Y is a compact subset of $G(\mathbb{A})$ and g ranges over Y, the function $h \mapsto W\left(\left(\begin{smallmatrix} h & 0 \\ 0 & 1 \end{smallmatrix}\right) g\right)$ is compactly supported on the space of $h \in N_{n-1}(\mathbb{A}) \backslash G_{n-1}(\mathbb{A})$, $|\det(h)| = 1$. Since $N_{n-1} \backslash G_{n-1}$ is discrete in this set, the sum $U(g)$ is finite. $\square$

Lemma 13.6. *For any $m \in \mathbb{Z}$, the restriction of $U''(h) = U\left(\left(\begin{smallmatrix} h & 0 \\ 0 & 1 \end{smallmatrix}\right)\right)$ to the set of $h \in G_{n-1} \backslash G_{n-1}(\mathbb{A})$, $|\det h| = q^m$ is compactly supported. It is 0 if m is large enough.*

Proof. By reduction theory for $G_{n-1}(\mathbb{A})$ there is a sequence $(m_v' \in \mathbb{Z}; v \in |X|)$ with $m_v' = 0$ for almost all v, and a compact subset $\Omega \subset N_{n-1}(\mathbb{A})$ such that

$G_{n-1}(\mathbb{A}) = G_{n-1} \cdot \mathfrak{S}((m'_v), \Omega)$, where the Siegel set $\mathfrak{S}((m'_v), \Omega)$ consists of the $h = uak$ in $G_{n-1}(\mathbb{A})$ with $u \in \Omega$, $k \in K = \prod_v K_v$ and $a = \text{diag}(a_1, \ldots, a_{n-1})$, $a_i \in \mathbb{A}^\times$, with $|a_i/a_{i+1}|_v \geq q^{m'_v}$ $(1 \leq i < n-1)$. It suffices to show that $U''(h)$ is compactly supported on $\mathfrak{S}_m((m'_v), \Omega) = \{h \in \mathfrak{S}((m'_v), \Omega); |\det h| = q^m\}$. It suffices to show that for h in $\mathfrak{S}_m$ with $U''(h) \neq 0$, $|a_1|$ is bounded. If $U''(h) \neq 0$ there is $p \in G_{n-1}$ with $W\left(\left(\begin{smallmatrix} ph & 0 \\ 0 & 1 \end{smallmatrix}\right)\right) \neq 0$. Suppose $p_{n-1,1} \neq 0$. Consider $ph = puak$. The $(n-1,1)$ entry of pua is $p_{n-1,1}a_1$. If $pua = bk'$, $k' \in K$ and $b \in B_{n-1}(\mathbb{A})$, then $|p_{n-1,1}a_1|_v \leq |b_{n-1,n-1}|_v$, and this is bounded by q^{m_v} by (1) of Lemma 4.1, for all v. Put $m' = \sum_v m_v$. Then $|a_1| = |p_{n-1,1}a_1| = \prod_v |p_{n-1,1}a_1|_v$ is bounded by $\prod_v q^{m_v} = q^{m'}$. If $p_{n-1,1} = 0$ we use the largest i with $p_{i,1} \neq 0$ to obtain $|p_{i,1}a_1|_v \leq |b_{i,i}|_v$ for all v. But $W\left(\left(\begin{smallmatrix} ph & 0 \\ 0 & 1 \end{smallmatrix}\right)\right) \neq 0$ implies that $b_{i,i}|_v \leq q^{m_v(n-i)}$. Hence $|a_1|$ is bounded by $q^{m'(n-i)}$. In particular, if $U''(h) \neq 0$, $|\det h| = \prod_i |a_i|$ $(1 \leq i < n)$ is bounded independently of m. $\qquad \square$

Write U_ξ for $U = U_W$ of Lemma 13.5(2) if $\pi \ni \xi \mapsto W_\xi \in W(\pi, \psi)$.

Lemma 13.7. *For any $g \in \text{GL}(n, \mathbb{A})$ with $W(g) \neq 0$ there exists $u \in N(\mathbb{A})$ with $U(ug) \neq 0$. Hence $U_\xi \not\equiv 0$ if $\xi \neq 0$.*

Proof. The ψ-Fourier coefficient of $U(g)$ is

$$\int_{N \backslash N(\mathbb{A})} U(ug)\overline{\psi}(u)du = \int_{N \backslash N(\mathbb{A})} \sum_{N \backslash P} W(pug)\overline{\psi}(u)du$$

$$= \int_{N \backslash N(\mathbb{A})} \sum_{h \in N_{n-1} \backslash G_{n-1}} W\left(\left(\begin{smallmatrix} h & 0 \\ 0 & 1 \end{smallmatrix}\right)ug\right)\overline{\psi}(u)du.$$

Let N'' be the unipotent radical of P (nonzero nondiagonal entries only in the last column). This N'' is normal in N. We may and do integrate over $N'' \backslash N''(\mathbb{A})$ first to get the sum over $h \in N_{n-1} \backslash G_{n-1}$ of

$$\int_{N_{n-1} \backslash N_{n-1}(\mathbb{A})} \int_{N'' \backslash N''(\mathbb{A})} W\left(\left(\begin{smallmatrix} h & 0 \\ 0 & 1 \end{smallmatrix}\right)u''ug\right)\overline{\psi}(u'')du''\overline{\psi}(u)du.$$

As G_{n-1} normalizes N'', putting h' for $\text{diag}(h,1)$, we have

$$W(h'u''ug) = \psi(h'u''h'^{-1})W(h'ug),$$

and $\int_{N'' \backslash N''(\mathbb{A})} \psi(h'u''h'^{-1}u''^{-1})du''$ is 1 if $h \in P_{n-1}$ and 0 if not. So the sum ranges only over $N_{n-1} \backslash P_{n-1}$. Continuing by induction on n we get $\int_{N \backslash N(\mathbb{A})} U(ug)\overline{\psi}(u)du = W(g)$. The first claim follows. For the second, since the map $\pi \ni \xi \mapsto W_\xi$ is injective, $U_\xi \not\equiv 0$. $\qquad \square$

Note that by construction U is left invariant under the parabolic $P' = ZP$ of type $(n-1,1)$. Moreover u is right invariant under an open subgroup of $G(\mathbb{A})$, since W is.

Write w_n for antidiag$(1, \ldots, 1)$ in G_n. Let α_n be $w_n \left(\begin{smallmatrix} w_{n-1} & 0 \\ 0 & 1 \end{smallmatrix}\right) = \left(\begin{smallmatrix} 0 & 1 \\ I_{n-1} & 0 \end{smallmatrix}\right)$. If $W = W_\xi$ lies in the Whittaker model of π, then $\widetilde{W}(g) = W(w_n \, {}^t g^{-1})$ lies in

the Whittaker model of the contragredient representation $\pi^\vee$ of π (see [BZ]).
Put

$$g' = \left(\begin{smallmatrix} w_{n-1} & 0 \\ 0 & 1 \end{smallmatrix}\right) {}^t g^{-1} \left(\begin{smallmatrix} w_{n-1} & 0 \\ 0 & 1 \end{smallmatrix}\right), \qquad \xi' = \pi^\vee \left(\left(\begin{smallmatrix} w_{n-1} & 0 \\ 0 & 1 \end{smallmatrix}\right)\right)\xi.$$

Define

$$V_\xi(g) = \sum_{p\in N_{n-1}\backslash G_{n-1}} \widetilde{W}_{\xi'}\left(\left(\begin{smallmatrix} p & 0 \\ 0 & 1 \end{smallmatrix}\right)g'\right).$$

For each $g \in \mathrm{GL}(n,\mathbb{A})$ this sum is finite. For each $m \in \mathbb{Z}$ the $h \in G_{n-1}\backslash G_{n-1}(\mathbb{A})$
with $V_\xi\left(\left(\begin{smallmatrix} h & 0 \\ 0 & 1 \end{smallmatrix}\right)\right) \neq 0$ and $|\det h| = q^m$ make a compact set, which is empty if
m is small enough, by Lemma 4.2. Note that if $N' = N'_n = \alpha_n^{-1} N_n \alpha_n$, then

$$V_\xi(g) = \sum_{p\in N'\backslash \overline{P}} W_{\xi'}(\alpha_n p g) = \sum_{P\in N_{n-1}\backslash G_{n-1}} W_{\xi'}\left(\alpha_n \left(\begin{smallmatrix} p & 0 \\ 0 & 1 \end{smallmatrix}\right)g\right).$$

Thus V_ξ is left invariant under the parabolic $\overline{P}' = {}^t P'$ opposite to P', and
right invariant under an open subgroup of $\mathrm{GL}(n,\mathbb{A})$.

Remark 13.1. The idea of the proof of the converse theorem is to recover
the following direct result: $U(g) = U(\gamma g) = V(g)$ for all $\gamma \in G_n$ and $g \in$
$G_n(\mathbb{A})$, if U and V are constructed from the Whittaker function $W(g) =$
$\int_{N_n\backslash N_n(\mathbb{A})} \varphi(ug)\overline{\psi}(u)du$ associated with a cusp form φ on $G_n\backslash G_n(\mathbb{A})$. The
equality $U(g) = V(g)$ would produce a function left invariant under P_n and
$\overline{P}_n$, hence under G_n, that is, automorphic. The direct result follows from
Fourier expansion inductively along the unipotent radicals of P_m, using the
assumption that φ is cuspidal: Put $d[\gamma_m] = \mathrm{diag}(\gamma_m, I_{n-m})$. Then

$$\begin{aligned}
\varphi(g) &= \sum_{\gamma_{n-1}\in P_{n-1}\backslash G_{n-1}} \cdots \sum_{\gamma_m\in P_m\backslash G_m} \cdots \sum_{\gamma_1\in G_1} W(d[\gamma_1]\ldots d[\gamma_{n-1}]g) \\
&= \sum_{\gamma\in N_{n-1}\backslash G_{n-1}} W(d[\gamma]g)=U(g), \text{ and } \varphi(g)=\sum_\gamma W(\mathrm{diag}(1,\gamma)\alpha_n g)=V(g).
\end{aligned}$$

13.4. **The Integrals I and Ψ.** Recall that for an idèle $a = (a_v)$ we put
$\deg((a_v)) = -\sum_v \deg(v)\deg_v(a_v)$, where $q_v = q^{\deg(v)}$ defines $\deg(v)$ and
$|a_v|_v = q_v^{-\deg_v(a_v)}$. Then $|a| = q^{\deg(a)}$. We also write $t = q^{-s}$ for $s \in \mathbb{C}$,
for comparison with the number field case. Then $|\det h|^{s-\frac{1}{2}}$, the factor which
appears over number fields, becomes $q^{-\frac{1}{2}\deg(h)}t^{-\deg(h)}$, where $\deg(h)$ means
$\deg(\det(h))$.

Let φ be an automorphic function on $\mathrm{GL}(n-1,\mathbb{A})$. Put

$$I(t,\xi,\varphi) = \int_{G_{n-1}\backslash G_{n-1}(\mathbb{A})} \varphi(h)U_\xi\left(\left(\begin{smallmatrix} h & 0 \\ 0 & 1 \end{smallmatrix}\right)\right) q^{-\frac{1}{2}\deg(h)}t^{-\deg(h)}dh,$$

$$\widetilde{I}(t,\xi,\varphi) = \int_{G_{n-1}\backslash G_{n-1}(\mathbb{A})} \varphi(h)V_\xi\left(\left(\begin{smallmatrix} h & 0 \\ 0 & 1 \end{smallmatrix}\right)\right) q^{-\frac{1}{2}\deg(h)}t^{-\deg(h)}dh.$$

These are well-defined formal Laurent series in t and t^{-1} by Lemma 4.2.

Using the definition of U_ξ and V_ξ as series in W_ξ and the property $W_\xi(ug) =$
$\psi(u)W_\xi(g)$ $(g \in \mathrm{GL}(n,\mathbb{A}), u \in N(\mathbb{A}))$, one computes

Lemma 13.8. *For h in $G_{n-1}(\mathbb{A})$ put $W_\varphi(h) = \int_{N_{n-1}\backslash N_{n-1}(\mathbb{A})} \varphi(uh)\psi(u)du$, $\widetilde{W}_\varphi(h) = W_\varphi(w_{n-1}{}^t h^{-1})$ and*

$$\Psi(t, W, W') = \int_{N_{n-1}(\mathbb{A})\backslash G_{n-1}(\mathbb{A})} W'(h)W\left(\left(\begin{smallmatrix} h & 0 \\ 0 & 1 \end{smallmatrix}\right)\right) q^{-\frac{1}{2}\deg(h)} t^{-\deg(h)} dh.$$

Then $I(t, \xi, \varphi) = \Psi(t, W_\xi, W_\varphi)$ and $\widetilde{I}(t, \xi, \varphi) = \Psi(1/qt, \widetilde{W}_\xi, \widetilde{W}_\varphi)$.

Lemma 13.9. *Let $\pi = \otimes_v \pi_v$ be a unitarizable irreducible admissible locally generic representation of $\mathrm{GL}(n, \mathbb{A})$ whose central character is trivial on $F^\times$. Let S_2 be a finite set of places of F. Suppose π' is an irreducible cuspidal representation of $\mathrm{GL}(n-1, \mathbb{A})$ whose components π'_v ($v \in S_2$) are unramified and such that the formal series $L(t, \pi \times \pi')$ and $L(t, \pi^\vee \times \pi'^\vee)$ are polynomials in t satisfying the functional equation*

$$L(t, \pi \times \pi') = \varepsilon(t, \pi \times \pi')L(1/qt, \pi^\vee \times \pi'^\vee).$$

Then for any vector $\xi = \prod_v \xi_v$ in $\pi = \otimes_v \pi_v$ and form φ in the space of π', we have $I(t, \xi, \varphi) = \widetilde{I}(t, \xi, \varphi)$.

Proof. By Lemma 13.8, it suffices to show that $\Psi(t, W_\xi, W_\varphi) = \Psi(1/qt, \widetilde{W}_\xi, \widetilde{W}_\varphi)$. The vector ξ in the abstract space $\pi = \otimes_v \pi_v$ can be taken to be a product, $\otimes_v \xi_v$. Hence $W_\xi((g_v)) = \prod_v W_{\xi_v}(g_v)$. The Whittaker function W_φ lies in the Whittaker model of $\pi' = \otimes_v \pi'_v$, which is the restricted product of the Whittaker models $W(\pi'_v, \overline{\psi}_v)$ of the components π'_v. Hence we may assume that W_φ has the form $W' = \prod_v W'_v$ (as W_φ is a finite linear combination of such functions). For such factorizable W_ξ and W' we have

$$\Psi(t, W_\xi, W') = \prod_v \Psi(t, W_{\xi_v}, W'_v), \qquad \Psi(t, \widetilde{W}_\xi, \widetilde{W}') = \prod_v \Psi(t, \widetilde{W}_{\xi_v}, \widetilde{W}'_v).$$

Taking the product over v of the local functional equations

$$\frac{\Psi(t, W_{\xi_v}, W'_v)}{L(t, \pi_v \times \pi'_v)}\varepsilon(t, \pi_v \times \pi'_v, \psi_v)\omega_{\pi'_v}(-1)^{n-1} = \frac{\Psi(1/qt, \widetilde{W}_{\xi_v}, \widetilde{W}'_v)}{L(1/qt, \pi_v^\vee \times \pi_v'^\vee)}$$

and using the functional equation $L(t, \pi \times \pi') = \varepsilon(t, \pi \times \pi')L(1/qt, \pi^\vee \times \pi'^\vee)$, the lemma follows. $\square$

Proposition 13.10. *Let S_1 be a finite set of places of F, disjoint from S_2. Suppose π is as in Lemma 13.9 and its components π_v are cuspidal at $v \in S_1$. Let $\xi = \otimes_v \xi_v$ be a vector in $\otimes_v \pi_v$ such that for each $v \in S_2$ the component ξ_v is $K'_v = \mathrm{GL}(n-1, R_v)$-invariant (and G_{n-1} embeds in G_n as usual). Suppose π' is as in Lemma 13.9 but its components π'_v ($v \in S_1$) are cuspidal. Then for any character ω of $F^\times \backslash \mathbb{A}^\times$ we have $U_\xi(e) = V_\xi(e)$.*

Proof. We have the equality $I(t, \xi, \varphi) = \widetilde{I}(t, \xi, \varphi)$ for any φ in any π' which is unramified at $v \in S_2$ and cuspidal at $v \in S_1$. The restriction at $v \in S_2$ means that the forms φ that we have consist of all those which are right invariant under a $G_{n-1}(F_{S_2})$-conjugate of $\prod_v K'_v$ ($v \in S_2$). So that I and $\widetilde{I}$ be nonzero we then need to take $\xi = \otimes_v \xi_v$ whose component ξ_v ($v \in S_2$) is K'_v-invariant (or a $G_{n-1}(F_{S_2})$-translate of it).

Let ω be a character of $F^\times \backslash \mathbb{A}^\times$, hence also of $Z_{n-1} \backslash Z_{n-1}(\mathbb{A})$. Suppose φ satisfies $\varphi(zg) = \omega(z)\varphi(g)$ $(g \in G_{n-1}(\mathbb{A}), z \in Z_{n-1}(\mathbb{A}))$. For $z \in F^\times \backslash \mathbb{A}^\times$ put $\omega_t(z) = \omega(z)(q^{-\frac{1}{2}}t^{-1})^{(n-1)\deg(z)}$ and

$$(A_{\omega_t} U_\xi)\left(\left(\begin{smallmatrix} h & 0 \\ 0 & 1 \end{smallmatrix}\right)\right) = \int_{Z_{n-1}\backslash Z_{n-1}(\mathbb{A})} U_\xi\left(\left(\begin{smallmatrix} zh & 0 \\ 0 & 1 \end{smallmatrix}\right)\right) \omega_t(z) dz.$$

It satisfies $(A_{\omega_t} U_\xi)\left(\left(\begin{smallmatrix} zh & 0 \\ 0 & 1 \end{smallmatrix}\right)\right) = \omega_t(z)^{-1}(A_{\omega_t} U_\xi)\left(\left(\begin{smallmatrix} h & 0 \\ 0 & 1 \end{smallmatrix}\right)\right)$ and

$$I(t,\xi,\varphi) = \int_{G_{n-1}Z_{n-1}(\mathbb{A})\backslash G_{n-1}(\mathbb{A})} \varphi(h)(A_{\omega_t} U_\xi)\left(\left(\begin{smallmatrix} h & 0 \\ 0 & 1 \end{smallmatrix}\right)\right)(q^{-\frac{1}{2}}t^{-1})^{\deg(h)} dh.$$

The map $\xi \mapsto U_\xi$ is $G_n(\mathbb{A})$-equivariant: $\pi(g)\xi \mapsto U_{\pi(g)\xi}$ and $U_{\pi(g)\xi}(h) = U_\xi(hg)$. In particular $(A_{\omega_t} U_\xi|G_{n-1}(\mathbb{A}))$ is an automorphic function on $G_{n-1}(\mathbb{A})$, and its restriction to $G_{n-1}(F_v)$, $v \in S_1$, lies in the space $\mathrm{ind}(G_{n-1,v}, Z_{n-1,v} N_{n-1,v}; \omega_{t,v}\psi_{n-1,v})$ by Corollary 13.4, since π_v is cuspidal for $v \in S_1$. Consequently $(A_{\omega_t} U_\xi|G_{n-1}(\mathbb{A}))$ lies in the space of cusp forms belonging to those cuspidal representations of $G_{n-1}(\mathbb{A})$ whose central character is ω_t^{-1}, their components at $v \in S_1$ are cuspidal, and their components at $v \in S_2$ are unramified. Since $I(t,\xi,\varphi) = \tilde{I}(t,\xi,\varphi)$ for any automorphic function φ on $G_{n-1}(\mathbb{A})$ which is cuspidal as a function on $G_{n-1,v}$ $(v \in S_1)$, unramified as a function on $G_{n-1,v}$ $(v \in S_2)$, with central character ω, we conclude that

$$\int_{Z_{n-1}\backslash Z_{n-1}(\mathbb{A})} [U_\xi\left(\left(\begin{smallmatrix} zh & 0 \\ 0 & 1 \end{smallmatrix}\right)\right) - V_\xi\left(\left(\begin{smallmatrix} zh & 0 \\ 0 & 1 \end{smallmatrix}\right)\right)]\omega_t(z) dz = 0.$$

This is a power series in t. Hence we conclude its coefficients are zero. Thus

$$\int_{Z_{n-1}\backslash Z_{n-1}^0(\mathbb{A})} [U_\xi\left(\left(\begin{smallmatrix} zh & 0 \\ 0 & 1 \end{smallmatrix}\right)\right) - V_\xi\left(\left(\begin{smallmatrix} zh & 0 \\ 0 & 1 \end{smallmatrix}\right)\right)]\omega(z) dz = 0$$

where $Z_{n-1}^0(\mathbb{A})$ is the subgroup of z in $Z_{n-1}(\mathbb{A})$ with $\deg(z) = 0$. Now ω ranges over the space of characters of the compact group $Z_{n-1}\backslash Z_{n-1}^0(\mathbb{A})$. Hence $U_\xi\left(\left(\begin{smallmatrix} h & 0 \\ 0 & 1 \end{smallmatrix}\right)\right) = V_\xi\left(\left(\begin{smallmatrix} h & 0 \\ 0 & 1 \end{smallmatrix}\right)\right)$ for all h in $G_{n-1}(\mathbb{A})$, and in particular $U_\xi(e) = V_\xi(e)$. $\quad\square$

13.5. Proof of the Simple Converse Theorem.

Recall that K_v denotes the maximal open subgroup $\mathrm{GL}(n, R_v)$ of $\mathrm{GL}(n, F_v)$, $v \in |X|$, and $K = \prod_v K_v$, a maximal compact subgroup of $\mathrm{GL}(n, \mathbb{A})$. For an integer $m_v \geq 0$ denote by $K_{1v}(m_v)$ the subgroup of g_v in K_v whose bottom row is $(0, \ldots, 0, 1) \bmod \pi_v^{m_v}$. In particular $K_{1v}(0)$ is K_v. Put $K_1((m_v)) = \prod_v K_{1v}(m_v)$ where $m_v \geq 0$ are integers for all v and $m_v = 0$ for almost all v.

Proposition 13.11. (1) *The parabolic subgroup $P_n'(F)$ of type $(n-1, 1)$ and its opposite parabolic $\overline{P}_n'(F) = {}^t P_n'(F)$ generate the discrete subgroup $\mathrm{GL}(n, F)$ in $\mathrm{GL}(n, \mathbb{A})$.* (2) *If $m_v > 0$ for some v then the subgroups $P_n'(F) \cap K_1((m_v))$ and $\overline{P}_n'(F) \cap K_1((m_v))$ generate the subgroup $\mathrm{GL}(n, F) \cap K_1((m_v))$ of $\mathrm{GL}(n, F)$.*

Proof. See Prop. 9.1 of [CPS], pp. 194–195. $\quad\square$

To prove the simple converse theorem we consider first the case where S_2 is empty. Then for every vector $\xi = \otimes_v \xi_v$ in $\pi = \otimes_v \pi_v$ we have $U_\xi(e) = V_\xi(e)$ by Prop. 13.10. Hence for every g in $\mathrm{GL}(n, \mathbb{A})$ we have $U_\xi(g) = U_{\pi(g)\xi}(e) =$

$V_{\pi(g)\xi}(e) = V_\xi(g)$, and so $U_\xi \equiv V_\xi$. This function is left invariant under $P'_n(F)$ and $\overline{P}'_n(F)$, hence by the group $\mathrm{GL}(n, F)$, by Prop. 13.11. It is right invariant under an open subgroup of $\mathrm{GL}(n, \mathbb{A})$, and it is cuspidal. The map $\xi \mapsto U_\xi$ is nonzero by Lemma 13.7. It is equivariant and defines a realization of the admissible irreducible representation $\pi = \otimes_v \pi_v$ of $\mathrm{GL}(n, \mathbb{A})$ in the space of automorphic cuspidal functions on $\mathrm{GL}(n, F)\backslash \mathrm{GL}(n, \mathbb{A})$.

When S_2 is not empty, we choose the vector ξ_v in the component π_v of π such that $W_{\xi_v}(e) = 1$ $(v \in S_2)$ and such that ξ_v is fixed by $K_{1v}(m_v)$, smallest $m_v \geq 0$. Such a vector ξ_v exists and is unique up to a scalar [JPS1]. In particular such ξ_v is fixed under the subgroup $\mathrm{GL}(n-1, R_v)$ of $K_{1v}(m_v) \subset K_v = \mathrm{GL}(n.R_v)$. Put $\xi_{S_2} = \otimes_{v \in S_2} \xi_v$.

By Prop. 13.10 for every vector $\xi^{S_2} = \otimes_{v \notin S_2} \xi_v$ in $\otimes_{v \notin S_2} \pi_v$, if $\xi = \xi^{S_2} \otimes \xi_{S_2}$ then U_ξ is equal to V_ξ at e, hence on the subgroup $\mathrm{GL}(n, \mathbb{A}^{S_2}) \cdot K_1((m_v; v \in S_2))$ of $\mathrm{GL}(n, \mathbb{A})$. Moreover this function is left invariant under both $P'_n(F) \cap K_1((m_v; v \in S_2))$ and $\overline{P}'_n(F) \cap K_1((m_v; v \in S_2))$, hence under the group $\mathrm{GL}(n, F) \cap K_1((m_v; v \in S_2)) = \mathrm{GL}(n, F) \cap \mathrm{GL}(n, \mathbb{A}^{S_2}) \cdot K_1((m_v; v \in S_2))$ by Prop. 13.11.

There exists a vector ξ^{S_2} such that $\xi = \xi^{S_2} \otimes \xi_{S_2}$ satisfies $W_\xi(e) \neq 0$, since $\xi \mapsto W_\xi$ is injective and $W_{\pi(g)\xi}(h) = W_\xi(hg)$. For such ξ^{S_2}, the restriction of U_ξ to $K_1((m_v; v \in S_2)) \mathrm{GL}(n, \mathbb{A}^{S_2})$ is nonzero. Indeed, U_ξ is left invariant under $P'_n(F)$, hence under $N_n(F)$. By strong approximation theorem

$$N_n(F) \cdot [N_n(\mathbb{A}) \cap \mathrm{GL}(n, \mathbb{A}^{S_2}) \cdot K_1((m_v; v \in S_2))]$$

is dense in $N_n(\mathbb{A})$ (since $F \cdot \mathbb{A}^{S_2} \cdot \prod_{v \in S_2} \pi_v^{m_v} R_v$ is dense in $\mathbb{A}$). By Lemma 13.7, $U_\xi(e) \neq 0$. Hence $U_\xi | \mathrm{GL}(n, \mathbb{A}^{S_2}) K_1((m_v; v \in S_2))$ is nonzero.

Define $U_{\xi^{S_2}}$ on $\mathrm{GL}(n, \mathbb{A})$ by $U_{\xi^{S_2}}(g) = U_\xi(g^0)$ $(\xi = \xi^{S_2} \otimes \xi_{S_2})$ if g has the form γg^0 with $\gamma \in \mathrm{GL}(n, F)$ and $g^0 \in \mathrm{GL}(n, \mathbb{A}^{S_2}) K_1((m_v; v \in S_2))$ (this definition is independent of the choice of γ and g^0 since U_ξ is left invariant under $\mathrm{GL}(n, F) \cap \mathrm{GL}(n, \mathbb{A}^{S_2}) K_1((m_v; v \in S_2)))$ and $U_{\xi^{S_2}}(g) = 0$ otherwise. The map $\xi^{S_2} \mapsto U_{\xi^{S_2}}$ is a nonzero equivariant homomorphism of the admissible irreducible representation $\otimes_{v \notin S_2} \pi_v$ of $\mathrm{GL}(n, \mathbb{A}^{S_2})$ into the space of right-smooth functions on $\mathrm{GL}(n, F) \backslash \mathrm{GL}(n, \mathbb{A})$ which transform under the center $Z_n(\mathbb{A})$ according to the character ω_π of π. The representation of $\mathrm{GL}(n, \mathbb{A})$ on the space generated by these $U_{\xi^{S_2}}$ is admissible since it is so for each $v \notin S_2$. It has an irreducible subrepresentation π'' which is necessarily automorphic, and its components at each $v \notin S_2$ are in the given π_v. In particular π'' is cuspidal.   □

Let F_u be a local field of positive characteristic. Put $G_u = \mathrm{GL}(n, F_u)$ and $G''_u = \mathrm{GL}(n-1, F_u)$. For any G_u-module π_u and G''_u-module τ_u we put

$$\Gamma(s, \pi_u, \tau_u) = \frac{L(s, \pi_u, \tau_u)}{\varepsilon(s, \pi_u, \tau_u)\tilde{L}(1-s, \pi_u, \tau_u)}.$$

Corollary 13.12. *Let π_u be a G_u-module such that there is a global field F whose completion at some place u is our F_u and a unitary irreducible cuspidal $G(\mathbb{A})$-module π whose component at u is our π_u. Let π'_u be an irreducible* *nondegenerate G_u-module such that for some set V of places of F not including u we have*

$$\Gamma(s, \pi_u, \tau_u) = \Gamma(s, \pi'_u, \tau_u)$$

for every G''_u-module τ_u which is the u-component of a unitary cuspidal G''-module τ with a cuspidal component at the places of S_1. Then π'_u is equivalent to π_u.

Proof. Put $\pi = \pi_u \otimes \pi^u$ and $\pi' = \pi'_u \otimes \pi^u$. Since π is unitary cuspidal, $L(s, \pi, \tau)$ is entire and satisfies the functional equation $\Gamma(s, \pi, \tau) = 1$ of Theorem 12.12 for all unitary cuspidal τ. Our assumption on π'_u implies that

$$\Gamma(s, \pi', \tau) = \Gamma(s, \pi, \tau)$$

for all τ in Φ. Hence Theorem 13.1 implies that there exists an automorphic $G(\mathbb{A})$-module $\pi'' = \otimes \pi''_v$ with $\pi''_u \simeq \pi'_u$ and $\pi''_v \simeq \pi_v$ for all $v \neq u$ outside V. The rigidity theory for $\mathrm{GL}(n)$ (see [JS]) implies that $\pi'' \simeq \pi$ since π is cuspidal and in particular $\pi_u \simeq \pi'_u$, as required. $\square$

Remark 13.2. Corollary 13.12 applies to any square-integrable G_u-module π_u. Indeed, given F_u there is a global F whose completion at some place u is F_u, and given π_u and a cuspidal π_w for some other place $w \neq u$ of F, it is easy to construct a cuspidal unitary $G(\mathbb{A})$-module π with these components π_u, π_w at u, w by means of the trace formula.

REFERENCES

[A1]   Arthur, J.: A trace formula for reductive groups I. Duke Math. J. **45**, 911–952 (1978)

[A2]   Arthur, J.: On a family of distributions obtained from orbits. Can. J. Math. **38**, 179–214 (1986)

[A3]   Arthur, J.: The local behaviour of weighted orbital integrals. Duke Math. J. **56**, 223–293 (1988)

[AM]   Atiyah, M., Macdonald, I.: Introduction to Commutative Algebra. Addison-Wesley, Reading (1969)

[B]   Bernstein, J.: P-invariant distributions on $GL(N)$. Lecture Notes in Mathematics **1041**, 50–102. Springer, New York (1984)

[BD]   Bernstein, J., rédigé par Deligne, P.: Le "centre" de Bernstein, dans Représentations des groupes réductifs sur un corps local. Hermann, Paris (1984)

[BDK]   Bernstein, J., Deligne, P., Kazhdan, D.: Trace Paley-Wiener theorem. J. Anal. Math. **47**, 180–192 (1986)

[BZ]   Bernstein, J., Zelevinski, A.: Representations of the group $GL(n, F)$ where F is a nonarchimedean local field. Uspekhi Mat. Nauk **31**, 5–70 (1976). (Russian Math. Surveys **31**, 1–68, 1976)

[Bo]   Borel, A.: Admissible representations of a semisimple group over a local field with vectors fixed under an Iwahori subgroup. Invent. Math. **35**, 233–259 (1976)

[BJ]   Borel, A., Jacquet, H.: Automorphic forms and automorphic representations. Proc. Sympos. Pure Math. **33**, I, 111–155 (1979)

[BN]   Bourbaki, N.: Commutative Algebra. Hermann, Paris (1972)

[C]   Casselman, W.: Characters and jacquet modules. Math. Ann. **230**, 101–105 (1977)

[CPS]   Cogdell, J., Piatetski-Shapiro, I.: Converse theorems for $GL(n)$. Publ. Math. Inst. Hautes Études Sci. **79**, 157–214 (1994)

[De1]   Deligne, P.: Formes modulaires et représentations de $GL(2)$. In: Deligne, P., Kuyk, W. (eds.) Modular Functions of One Variable II. Antwerpen Conference 1972, Springer Lecture Notes, vol. **349**, pp. 55–105. Springer, New York (1973)

[De2]   Deligne, P.: Les constantes des équations fonctionnelles des fonctions L. Lecture Notes in Mathematics **349**, 501–597. Springer, New York (1973). http://www.springerlink.com/content/t5v714531j557n02/fulltext.pdf

[De3]   Deligne, P.: La conjecture de Weil: II. Publ. Math. IHES **52**, 137–252 (1980)

Y.Z. Flicker, *Drinfeld Moduli Schemes and Automorphic Forms: The Theory*   149
of Elliptic Modules with Applications, SpringerBriefs in Mathematics,
DOI 10.1007/978-1-4614-5888-3, © Yuval Z. Flicker 2013

[DF]   Deligne, P., Flicker, Y.: Counting local systems with principal unipotent local monodromy. Annals of Math. (2013). http://www.math.osu.edu/~flicker.1/df.pdf

[DH]   Deligne, P., Husemoller, D.: Survey of drinfeld modules. Current Trends in Arithmetical Algebraic Geometry (Arcata, Calif., 1985), pp. 25–91. Contemporary Mathematics, vol. 67, American Mathematical Society, Providence (1987)

[D1]   Drinfeld, V.: Elliptic modules. Mat. Sbornik **94** (136) (1974)(4)= Math. USSR Sbornik **23** (1974), 561–592.

[D2]   Drinfeld, V.: Elliptic modules. II. Mat. Sbornik **102** (144) (1977)(2)= Math. USSR Sbornik **31** (1977), 159–170.

[F1]   Flicker, Y.: The trace formula and base change for GL(3). In: Lecture Notes in Mathematics, vol. 927. Springer, New York (1982)

[F2]   Flicker, Y.: Rigidity for automorphic forms. J. Anal. Math. **49**, 135–202 (1987)

[F3]   Flicker, Y.: Regular trace formula and base change lifting. Am. J. Math. **110**, 739–764 (1988)

[F4]   Flicker, Y.: Base change trace identity for U(3). J. Anal. Math. **52**, 39–52 (1989)

[F5]   Flicker, Y.: Regular trace formula and base change for GL(n). Ann. Inst. Fourier **40**, 1–36 (1990)

[F6]   Flicker, Y.: Transfer of orbital integrals and division algebras. J. Ramanujan Math. Soc. **5**, 107–121 (1990)

[F7]   Flicker, Y.: The tame algebra. J. Lie Theor. **21**, 469–489 (2011)

[FK1]   Flicker, Y., Kazhdan, D.: Metaplectic correspondence. Publ. Math. IHES **64**, 53–110 (1987)

[FK2]   Flicker, Y., Kazhdan, D.: A simple trace formula. J. Anal. Math. **50**, 189–200 (1988)

[FK3]   Flicker, Y., Kazhdan, D.: Geometric Ramanujan conjecture and Drinfeld reciprocity law. In: Number Theory, Trace Formulas and Discrete subgroups. In: Proceedings of Selberg Symposium, Oslo, June 1987, pp. 201–218. Academic Press, Boston (1989)

[Fu]   Fujiwara, K.: Rigid geometry, Lefschetz-Verdier trace formula and Deligne's conjecture. Invent. math. **127**, 489–533 (1997)

[GK]   Gelfand, I., Kazhdan, D.: On representations of the group GL(n, K), where K is a local field. In: Lie Groups and Their Representations, pp. 95–118. Wiley, London (1975)

[H]   Henniart, G.: Caractérisation de la correspondance de Langlands locale par les facteurs ε de paires. Invent. Math. **113**, 339–350 (1993)

[JPS1]   Jacquet, H., Piatetskii-Shapiro, I., Shalika, J.: Conducteur des représentations du groupe linéaire. Math. Ann. **256**, 199–214 (1981)

[JPS]   Jacquet, H., Piatetski-Shapiro, I., Shalika, J.: Rankin-Selberg convolutions. Am. J. Math. **104**, 367–464 (1982)

[JS]   Jacquet, H., Shalika, J.: On Euler products and the classification of automorphic forms II. Am. J. Math. **103**, 777–815 (1981)

[K1]   Kazhdan, D.: Cuspidal geometry of p-adic groups. J. Anal. Math. **47**, 1–36 (1986)

[K2]   Kazhdan, D.: Representations of groups over close local fields. J. Anal. Math. **47**, 175–179 (1986)

[Ko]   Koblitz, N.: p-adic numbers, p-adic analysis, and zeta functions, 2nd edn., GTM, vol. 58. Springer, New York (1984)

[Lf1]   Lafforgue, L.: Chtoucas de Drinfeld et conjecture de Ramanujan-Petersson. Asterisque **243**, ii+329 (1997)

[Lf2]   Lafforgue, L.: Chtoucas de Drinfeld et correspondance de Langlands. Invent. Math. **147**, 1–241 (2002)

[Lm1]   Laumon, G.: Transformation de Fourier, constantes d'équations fonctionelles et conjecture de Weil. Publ. Math. IHES **65**, 131–210 (1987)

[Lm2]   Laumon, G.: Cohomology of Drinfeld Modular Varieties, volumes I et II. Cambridge University Press, Cambridge (1996)

[LRS]   Laumon, G., Rapoport, M., Stuhler, U.: $\mathcal{D}$-elliptic sheaves and the Langlands correspondence. Invent. math. **113**, 217–338 (1993)

[Mi]   Milne, J.: Étale cohomology, Princeton Mathematical Series, vol. **33**. Princeton University Press, Princeton (1980)

[P]   Pink, R.: On the calculation of local terms in the Lefschetz-Verdier trace formula and its application to a conjecture of Deligne. Ann. Math. **135**, 483–525 (1992)

[S1]   Serre, J.P.: Zeta and L-functions. In: Schilling, O.F.G. (ed.) Arithmetic Algebraic Geometry. Proc. Conf. Purdue University, 1963. Harper and Row, New York (1965)

[S2]   Serre, J.P.: Abelian ℓ-adic Representations and Elliptic Curves. Benjamin, New-York (1968)

[Sh]   Shintani, T.: On an explicit formula for class 1 "Whittaker functions" on GL_n over p-adic fields. Proc. Japan Acad. **52**, 180–182 (1976)

[Sp]   Shpiz, E.: Thesis. Harvard University, Cambridge (1990)

[Ta]   Tate, J.: p-divisible groups. In: Proceedings of Conference on Local Fields, NUFFIC Summer School, Driebergen, Springer (1967)

[V]   Varshavsky, Y.: Lefschetz-Verdier trace formula and a generalization of a theorem of Fujiwara. Geom. Funct. Anal. **17**, 271–319 (2007)

[W]   Waterhouse, W.: Introduction to affine group schemes, GTM **66**. Springer, New York (1979)

[Z]   Zelevinski, A.: Induced representations of reductive p-adic groups II. On irreducible representations of $GL(n)$. Ann. Scient. Ec. Norm. Sup. **13**, 165–210 (1980)

[Zi]   Zink, Th: The Lefschetz trace formula for an open algebraic
            surface. In: Automorphic Forms, Shimura Varieties, and *L*-
            Functions, vol. II (Ann Arbor, MI, 1988), pp. 337–376. Perspec-
            tives of Mathematics, vol. 11, Academic Press, Boston (1990)
[EGA]   Grothendieck, A., Dieudonné, J.: Éléments de géométrie
            algébrique. Springer, Berlin (1971)
[SGA1]   Grothendieck, A.: Revêtements étales et groupe fondamental.
            In: Lecture Notes in Mathematics, vol. **224**. Springer, New York
            (1971)
[SGA4]   Artin, M., Grothendieck, A., Verdier, J.-L.: Théorie des topos et
            cohomologie étale des schémas. In: Lecture Notes in Mathematics,
            vol. **269, 270, 305**. Springer, New York (1972–1973)
[SGA4 1/2] Deligne, P.: Cohomologie étale. In: Lecture Notes in Mathemat-
            ics, vol. **569**. Springer, New York (1977)
[SGA5]   Grothendieck, A.: Cohomologie ℓ-adique et fonctions *L*. In: Lec-
            ture Notes in Mathematics, vol. **589**. Springer, New York (1977)

INDEX